Man and Forest Series
No. 4

General Editors

Klaus Seeland
Franz Schmithüsen

A joint research venture of the

Council of Cultural Growth and Cultural Relations, Cuttack

and

Chair of Forest Policy and Forest Economics,
Swiss Federal Institute of Technology (ETH), Zurich

in co-operation with

Deutsche Gesellschaft für Technische Zusammenarbeit (GTZ)
Tropenökologisches Begleitprogramm (TOEB)
(Tropical Ecology Support Programme)

Man and Forest Series

General Editors: Klaus Seeland; Franz Schmithüsen

The Man and Forest Series consists of monographs and readers on the use and management of forests and an account of forest lifestyles in various regions of South Asia. It aims at providing the findings of an eight years research project (1992-99) to the public. The volumes in this series are the outcome of a cross-cultural research network between the Chair of Forest Policy and Forest Economics, Swiss Federal Institute of Technology (ETH), the Deutsche Gesellschaft für Technische Zusammenarbeit (GTZ) and Non-Government as well as Government Organisations in India, Nepal and Bhutan. The series welcomes contributions covering relevant issues in the man-forest interface.

Essential Forest Produce in Orissa

Essential Forest Produce in Orissa

by

Nityananda Patnaik

D.K. Printworld (P) Ltd.
New Delhi

Cataloging in Publication Data — DK

Patnaik, N. (Nityananda), 1927 -

Essential forest produce in Orissa.
(Man and forest series; no. 4).
Includes bibliographical references (p.)
Includes index.

1. Forest products – India – Orissa. 2. Forests and forestry – Economic aspects – India – Orissa. I. Title. II. Series: Man and forest series; no. 4.

ISBN 81-246-0208-5

First published in India in 2003

Supported by: Tropical Ecology Support Programme (TOEB)
Deutsche Gesellschaft für Technische Zusammenarbeit (GTZ)

Published and printed by:
D.K. Printworld (P) Ltd.
Regd. office : '*Sri Kunj*', F-52, Bali Nagar
New Delhi - 110 015
Phones : (011) 545-3975; 546-6019
Fax : (011) 546-5926
E-mail: dkprintworld@vsnl.net
Web: www.dkprint.com

Foreword

The decision to publish Dr. Patnaik's comprehensive study on Non-Timber Forest Produce in Orissa in the series *Man & Forest* was taken because it provides a lot of valuable information relating to this important sector of the Orissan economy. The data on this subject are scattered and only few official publications are available. The question, how reliable these data are is, however, to be left open. Even insiders and experts are not always in a position to give an appropriate assessment neither of their accuracy nor their political value. Despite that it is a merit which is not to be underrated that the author has devoted much effort to the laborious task of data hunting and compilation. We are well aware of the fact that they are highly prone to become outdated within a short period of time. Yet, to have an overview on the economic situation of India's tribal population based on statistical evidence is maybe better than to rely only on estimations.

Another reason for the publication of this study in the series *Man & Forest* is the plea for a paradigmatic change in view of, as it is still officially called in India, Minor Forest Produce. The author's argumentation in this publication favours a change for a political recognition of all Non-Timber Forest Produce and an encouragement to acknowledge its vital economic and cultural importance for all forest dwelling tribal groups in India, who have to make their living predominantly from the forest. We therefore agreed with the author to change the term Minor Forest Produce into Essential Forest Produce wherever it was not used in an official document or context.

We do hope that by renaming a well introduced term, the policy discourse on the socio-economic value and cultural importance of the products referred to in this book will be inspiring to the reader. It may be taken as a well wishing attempt to draw the attention of concerned people to an important sector of produce that long since failed to be properly understood by many forest professionals. As a consequence, the wider public also tended to ignore what is essential for a tribal life-style, because it has not been adequately informed about it.

The focus on the socio-economic and cultural relevance of Essential Forest Produce is a new aspect in Indian forest policy. Conservation of biodiversity of flora and fauna has become an important issue during the recent decades. Both are increasingly taken serious in the public and professional forest policy discourse. It is more and more accepted that biodiversity can be safeguarded best if a forest is managed in an encompassing way with a wide range of varying moderate use and extraction of renewable resources. Where commercial over-exploitation of single commodities such as value timber, bamboo and certain other raw materials fetching high prices failed to be sustainable in the long run, tribal everyday forest management largely was, wherever possible.

This is mainly due to the dependence of the tribal population on local resources which is the underlying principle of any subsistence economy. If tribal groups are over-exploiting certain items of Essential Forest Produce, it is generally a result of wage harvesting for industrial purposes. In the context of procurement of local raw materials and rural employment opportunities Essential Forest Produce matters as a source of income generation. More sustainable, however, is a broad resource basis for a self-sufficient livelihood of the tribal communities which is based on a sound diversity of plant and animal species. Some of the political decision makers, anthropologists, foresters and those administrators who are concerned with the management of Essential Forest Produce may know this already, but all others are welcome to venture into this little known world of Orissa.

Klaus Seeland
Franz Schmithüsen

Preface

THE main objectives of the study on EFP are :

1. To make a list of EFP being collected by the tribals, quantity of collection, and economic importance of different items of EFP.
2. To give a sketch of marketing network and determine the degree of elimination of exploitation of tribals by the traders, middlemen and moneylenders.
3. To study the laws, rules and regulations in respect of forests and the role played by different co-operative organisations, corporations and Government of Orissa undertaking in protecting the tribal communities against exploitation.
4. To find out existing situation in regard to processing of different items of EFP and assess the prospects of value addition to benefit the tribal people and to explore the possibilities of processing of EFP items at the tribal level.
5. To review credit packages and products to tribals as well as to the procurement, marketing and processing agencies.
6. Finally to recommend (a) changes, if any required in the existing legal provisions, to prevent exploitation and ensure remunerative price to tribal people; (b) to suggest measures for improvement in the functioning of corporations, co-operative societies and Government of Orissa undertakings connected with procurement and marketing of EFP; (c) to assess credit packages for tribals with suggestions for processing on a priority basis and (e) to prepare a few bankable/viable projects.

Keeping these objectives in view, the study was undertaken in

Dhenkanal and Karanjia divisions of Angul Circle, Parlakhemundi, Phulbani and Puri divisions of Berhampur circle, Rayagada and Kalahandi of Koraput circle, Bonai and Rairakhol divisions of Sambalpur circle. The investigating team collected data in respect of collection of different items of EFP, quantity collected and the authorised agencies in charge of procurement of EFP. They also gathered information regarding payments due to tribal collectors, role of middlemen, method of collection, price fixation, storage, transport system, forest-based industries and marketing organisations.

In addition to the fieldwork through which primary data were collected by interviewing persons concerned with EFP, data were gathered from secondary sources from the office of PCCF, OFDC, TDCC, UFP, ORMAS, TRIFED, IPICOL and AMCS, Tikabali and a few LAMPS. A few weekly markets were also observed to find out business transactions in respect of EFP between the procuring agencies and tribal communities.

The study was scheduled to be completed in four months. The work started in October, 1995 and went on till the middle of March, 1996. More than one month over the stipulated time period was needed for the completion of the project. Moreover, compared with the extensive and complex nature of the study the time limit was very short and therefore full justice could not be given to make it more comprehensive.

Multiple agencies, both official and non-official are involved in the procurement of EFP. To ascertain real quantum of various items procured is an impossibility. Clandestine collection of EFP by private traders is difficult to ascertain and therefore are off the record. Branches and sub-offices of the corporations are located all over the State. To reach these centres for collection of data need money and time which were in short supply with the project. A study on EFP is largely a technical affair. Requisite knowledge in Botany, Chemistry, Climatology, Physiography and Business Organisations is necessary. Scarcity of fund rendered it impossible for consultation with specialists. Moreover, the time constraint stood in the way of giving training to the investigating team on the Chemistry and Botany of various items of EFP. Record-keeping at various levels is not systematic and therefore it required much longer time for collection of data. Persons dealing with EFP in various organisations were found busy in multiple works which were assigned to them to comply with time limit. They could not give sufficient time to us for collection of data to our satisfaction. Nevertheless the study was completed in spite of these limitations.

The study has brought out some very important findings. One of the findings is that there are innumerable items of EFPs unexplored in regard to their productive potentialities and utility. The tribal communities are living in the unestimable treasure and depend primarily on this treasure to meet their daily needs. Compared with the vastness of these gifts of nature the quantum of utilisation of these resources by the tribal people is just like a drop in the ocean. Through various five year plans and special programmes and projects the tribal communities are being helped to bring them up above the poverty line and remove other disabilities they are suffering from. But no dent has been made to eradicate poverty, illiteracy, squalor, ill-health and malnutrition in which they are steeped since time beyond memory. It is paradox that poverty exists in plenty. Each and every item of EFP whether it is fibre or a bark or a seed or even a kind of grass has immeasurable potentialities for improving the living conditions of those who are born and brought up in their midst. An inventory of this rich treasure assumes a matter of greatest importance and should be taken up on priority basis. Next comes exploration of processing of EFP at the tribal level for value addition to the tribal communities. Works on utility of various products both processed and unprocessed form should be taken up next in the priority of exploratory programme on EFP.

Nityananda Patnaik

Contents

Acknowledgements

IN course of my discussions with Shri K.G. Karmakar, General Manager, NABARD, Bhubaneswar regarding various problems relating to the tribes of Orissa and their development I was inspired by some of his innovative ideas and deeper insight into the prospects of tribal development. His Head-Office at Bombay was interested in taking up a study on EFP in Orissa with particular reference to processing of these items for value addition at the tribal level. He wrote to me a letter to find out if I was interested in taking up the study. I expressed my willingness and submitted a proposal. After scrutiny I was finally entrusted with this study. This is the genesis of the study and I remain ever grateful to Shri Karmakar for associating me in this little known but very important and extremely fascinating subject. Shri D.P. Panda, Deputy General Manager, was equally interested in this study and he along with Shri J. Maharana, Manager of NABARD extended all possible help to me for carrying out this study.

Shri M.C. Das, I.F.S. (Retd.) deserves special thanks for technical guidance.

I would like to thank the Assistant General Managers, NABARD, of different districts who helped the investigating team during their visit for collection of data from various procuring and marketing agencies.

Special thanks are also due to Shri A.P. Padhy, I.A.S., Managing Director, IPICOL, Shri N.C. Patnaik, I.F.S., PCCF, Shri Bhaskar Mahapatra, I.F.S., PCCF, Shri Bhaskar Mahapatra, I.FS., Managing Director, OFDC, Shri A.D. Mishra, Regional Manager, TRIFED, Shri R.C. Patro (O.A.S.-1), Managing Director, TDCC, Shri R.N. Rath, General Manager, IPICOL, Shri L.R. Behera, I.F.S., Marketing Manager (KL), OFDC, Shri Lalmohan Baske, Marketing Manager, TDCC, Shri Muralidhar Panda, Industry Officer, OFDC, Smt. Seema Pati, Deputy Chief Executive, ORMAS, Shri H.O. Saxena, Scientist,

RRL, Bhubaneswar for providing data from their respective organisations on various aspects of the study.

Thanks are also due to the Investigating team consisting of Shri B.N. Rout, Shri D. Raipitam, Shri B. Harichandan, Shri N.C. Rout and Shri D. Ojha.

Credit goes to Smt. Kiran Bala Devi for collection of data from various offices and corporations, like OFDC, TDCC, TRIFED, IPICOL, CCF (KL), PCCF and her valuable help in various stages of report writing and editing.

To record help and co-operation in data collection two names which are worthy of note are Shri Premchand Patnaik, Section Officer, PCCF and Shri Manorajan Mohanty, Marketing Supervisor, TDCC. They have taken pains in sparing their valuable time in discussions with us on various aspects of EFP.

For logistic support and co-ordination I appreciate the help of Shri Manoranjan Patnaik.

The data processing was done in the computer section of Social Science and Development Research Institute (SSADRI). The overall supervision and credit of computerisation, layout design and format of the report goes to Mr. Pradeep Satapathy and Smt. Nilanjana Das. The Stenographers Shri Sisir Kumar Patra and Shri F. Samal made untiring efforts in getting the drafts typed over and over again till every part of the work was in order. They deserve thanks for their labour.

I approached Prof. (Dr.) Klaus Seeland and Prof. (Dr.) Schmithüsen of Swiss Federal Institute of Technology, ETH Zentrum, Zurich, Switzerland if they would be interested for publication of my work on Essential Forest Produce in Orissa in their series MAN AND FOREST which D.K. Printworld is publishing. I am thankful to both of them for their decision to publish my book as No. 4 in the series. I am indebted to them for their favourable decision in this matter.

List of Abbreviations

AMCS	:	Agency Marketing Co-operative Society
CCF	:	Chief Conservator of Forest
CCRIMH	:	Central Council of Research in Indian Medicines and Homoeopathy
CIMPO	:	Central Indian Medicinal Plant Organisation
DM	:	Divisional Manager
DPF	:	Demarcated Protected Forest
DWCRA	:	Development of Women and Children in Rural Areas
EFP	:	Essential Forest Produce
FD	:	Forest Division
FRI	:	Forest Research Institute
ICSIR	:	Indian Council of Scientific and Industrial Research
IPICOL	:	Industrial Promotion and Investment Corporation Ltd.
KL	:	Kendu Leaf
KLCC	:	Kendu Leaf Co-ordinating Committee
KVIC	:	Khadi and Village Industries Corporation
LAMPS	:	Large Size Multipurpose Co-operative Society
NAFED	:	National Agricultural Co-operative Marketing Federation of India Ltd.
LRI	:	Lac Research Institute
MFP	:	Minor Forest Produce
NA	:	Not available
NABARD	:	National Bank for Agricultural and Rural Development

NBG	:	National Botanical Garden
NCA	:	National Commission on Agriculture
NCDBA	:	National Committee on Development of Backward Areas
NCDC	:	National Co-operative Development Corporation
OFDC	:	Orissa Forest Development Corporation Ltd.
ORMAS	:	Orissa Rural Development and Marketing Society
OSCB	:	Orissa State Co-operative Bank
OSCMF	:	Orissa State Co-operative Marketing Federation
OSFC	:	Orissa State Financing Corporation
PCCF	:	Principal Chief Conservator of Forest
PFC	:	Price Fixation Committee
RCS	:	Registrar, Co-operative Societies
R&D	:	Research and Development
RF	:	Reserve Forest
RMCS	:	Regional Marketing Co-operative Society
RRL	:	Regional Research Laboratory
RM	:	Regional Manager
SAP	:	Surplus Agricultural Produce
SBI	:	State Bank of India
SC	:	Scheduled Caste
SFRI	:	Social Forest Research Institute
ST	:	Scheduled Tribe
TDCC	:	Tribal Development Co-operative Corporation of Orissa Ltd.
TRIFED	:	Tribal Co-operative Marketing Development Federation of India, Ltd.
TSPA	:	Tribal Sub-Plan Area
UCF	:	Unclassified Forest
UDPF	:	Undemarcated Protected Forest
UFP	:	Utkal Forest Product Ltd.

List of Tables

List of Visuals

List of Annexures

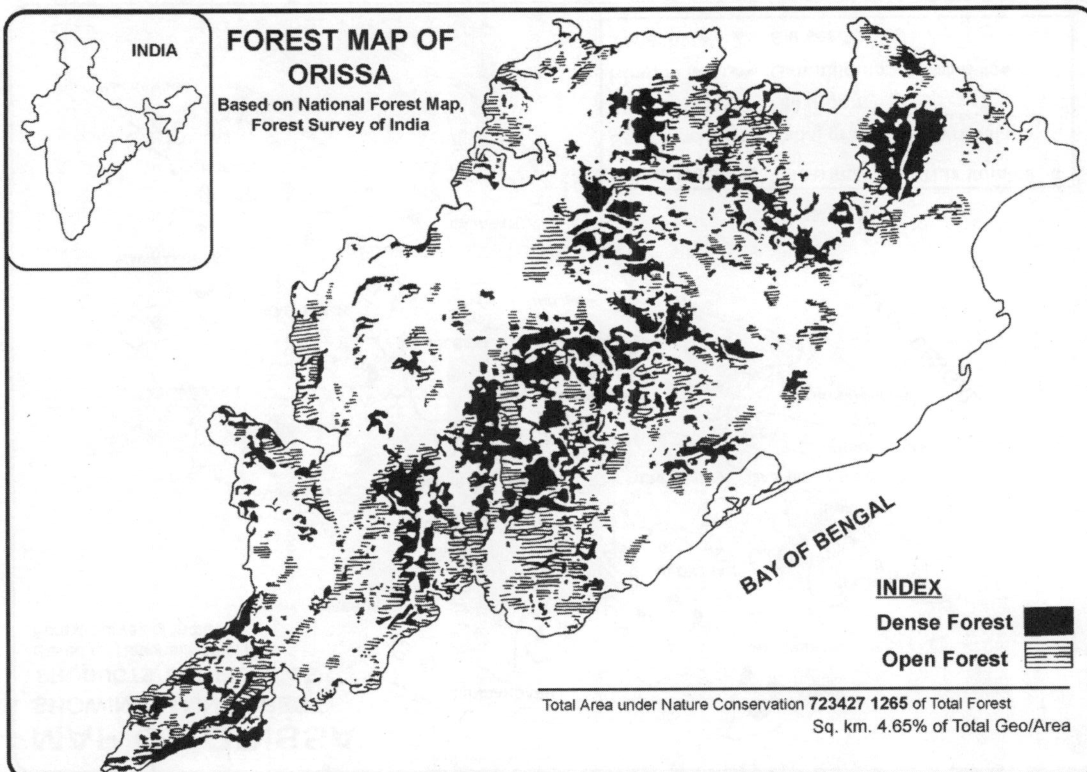
INDIA
FOREST MAP OF ORISSA
Based on National Forest Map, Forest Survey of India
BAY OF BENGAL
INDEX
Dense Forest
Open Forest
Total Area under Nature Conservation 723427 1265 of Total Forest
Sq. km. 4.65% of Total Geo/Area

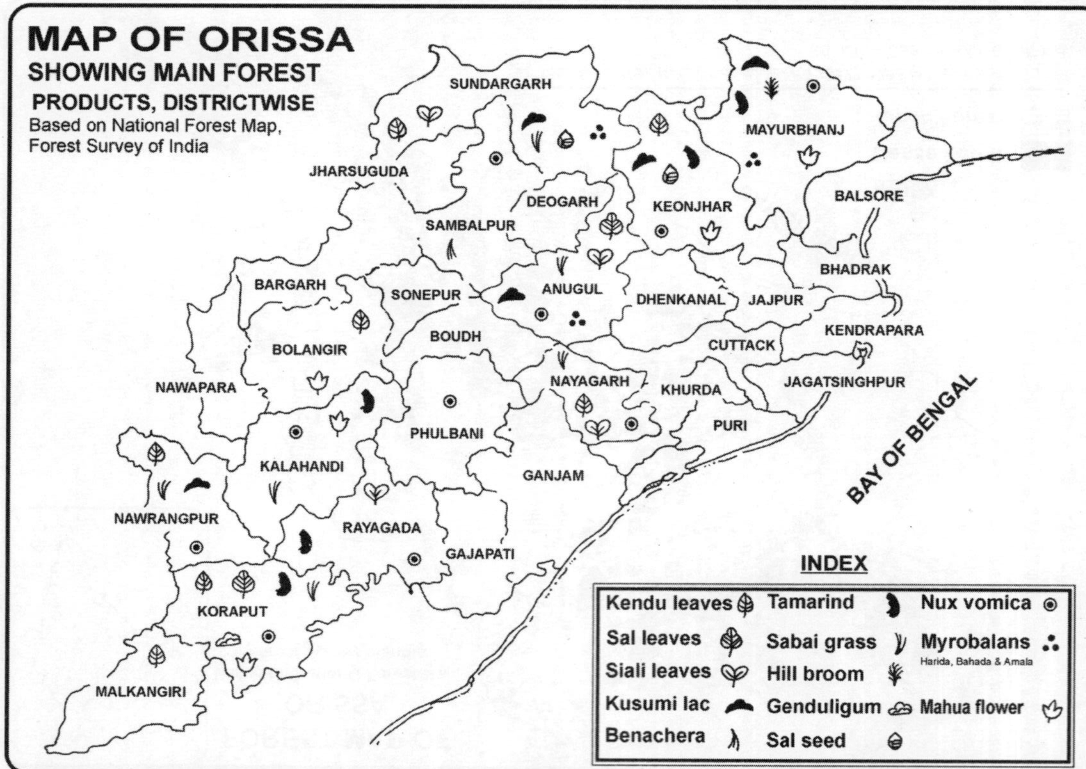
MAP OF ORISSA
SHOWING MAIN FOREST
PRODUCTS, DISTRICTWISE
Based on National Forest Map,
Forest Survey of India
SUNDARGARH
MAYURBHANJ
JHARSUGUDA
BALSORE
DEOGARH
KEONJHAR
SAMBALPUR
BHADRAK
BARGARH
SONEPUR
ANUGUL
DHENKANAL
JAJPUR
KENDRAPARA
BOUDH
BOLANGIR
CUTTACK
JAGATSINGHPUR
NAWAPARA
NAYAGARH
KHURDA
PURI
PHULBANI
KALAHANDI
GANJAM
BAY OF BENGAL
NAWRANGPUR
RAYAGADA
GAJAPATI
KORAPUT
MALKANGIRI
INDEX
Kendu leaves
Tamarind
Nux vomica
Sal leaves
Sabai grass
Myrobalans
Harida, Bahada & Amala
Siali leaves
Hill broom
Kusumi lac
Genduligum
Mahua flower
Benachera
Sal seed

1

Theme of the Study

Essential Forest Produce : Definition and scope

Orissa is very rich in Essential Forest Produce (EFP). The National Commission on Agriculture has defined Essential/Minor Forest Produce (EFP/MFP) as 'all produce obtainable from forest other than wood' (NCA, 1976, p. 229). The Orissa Timber and Other Forest Produces Transit Rules, 1980 has defined MFP as 'the produce other than timber, firewood, charcoal and bamboo'.

There are more than 100 items which can come under this category. Different items of EFP/MFP have the potential to bring about an economic revolution for the tribals and other rural people of the country. During drought and adverse climatic conditions the tribals depend mainly on EFP/MFP collection for their sustenance. A large number of items are not collected and lie waste in different forest regions because some of these areas are difficult of access and the tribals do not find their collection remunerative.

A big percentage of tribal population is at present being forced to eke out their subsistence by cultivation of marginal lands, shifting cultivation, overgrazing of forest areas and exploitation of trees. It is necessary to provide economically viable schemes of EFP/MFP collection, processing and marketing which can ensure tribals a reasonable sustenance and processing can employ not only able-bodied workers but also infirms, women and children in their respective habitats.

Area under forest

Over a geographical area of about 1,55,707 sq km, forests occupy 57,183.37 sq km (36.72 per cent) in Orissa. The break-up of the forest areas is as follows (as per 1990-1 figures) :

Reserve Forest	27,087.34 sq km
Demarcated Protected Forest	16,113.33 sq km
Undemarcated Protected Forest	13,966.97 sq km
Undemarcated Community Forest (UCF)	15.93 sq km

Per capita forest area in Orissa records at 0.18 ha. The forest revenue per ha. in Orissa comes to Rs. 103.49. The total revenue (1988-9) to the state exchequer was Rs. 59,18,19,000.00. It shows that forests are the most important source of revenue in the State. The State symbols in respect of birds, animals and trees are *Haliastur indus* (Sankachila-Golden Kites), *Cervus unicolor* (Sambar) and *Anogeissus acuminata* (Roxb. ex DC.) Guill & Perr. (Fasi) respectively.

The forest area differs from one territorial division to the other. Of all the divisions Rayagada division records highest forest area, that is, 5,019.73 sq km. Next comes Parlakhemundi which records to the extent of 4,368.63 sq km. followed by Kalahandi recording the forest area at 3,850.22 sq km. There are some divisions such as Baliguda, Keonjhar, Jeypore with forest areas varying from 2000 to 3000 sq. km. But there are many divisions which fall within the range from 1000 to 2000 sq km in extent. The divisions which record very low forest area are Ghumsar North and Karanjia with forest area below 1000 sq km. Annexure 1 gives forest areas of different territorial divisions as on 01.04.90 by legal status. So far as districts are concerned, Koraput and Ganjam district (old dist.) recorded the highest forest area which covers 8,866.79 sq km followed by Ganjam district of 7,032.88 sq km. Kalahandi 5,653.59 sq. km., Phulbani 5,651.10 sq km and Mayurbhanj 5,256.66 sq km. The district which records lowest is Balasore, that is, 258.13 sq km. Table 1.1 gives the forest types of Orissa and Table 1.2 gives the districtwise distribution of forest area *vis-a-vis* geographical area, population, percentage of forest area and per capita forest area.

Types of forests

The classification of forests as per Champion is as follows. Of all the types the major types are (a) Tropical, (b) Montane Sub-Tropical, (c) Montane Temperate, and (d) Alpine and specialised forest. Of all these tropical forests and specialised local types dominate in Orissa.

Table 1.1*: Forest Types of Orissa

Name of the Agroclimatic zone	*Types of forest as per Champion and Seth classification*
North-Central Plateau	North tropical Moist deciduous forest valley • Sal forest • Moist Sal Savannah • High level Sal forests
North-Western Plateau	• Dry peninsular Sal forests • Valley Sal forest (Moist Peninsular Sal Forests)
Western Undulating Plains	• Dry mixed deciduous forests • Mixed deciduous forest • Dry teak forest
Undulating Plains of Eastern Ghats	• Dry deciduous scrub • Seen in almost all • Dry teak forest • Seen in almost all • Seral type of the moist decduous forest • Moist peninsular Sal (high level)
Eastern Ghats Southern Uplands	• Southern tropical moist deciduous forest (teak forest) • Southern Moist mixed deciduous forest • High level moist peninsular (High level)
South-Eastern Region of Eastern Coastal Plains	• Low level moist peninsular Sal Forest • Tropical semi-average forests • Valley Sal • I. tomentoss seral type of moist deciduous forest
East & South Eastern Coastal Plains	• Mixed deciduous forest • I. tomentosa seral Coastal Plains type of moist deciduous forest • Tropical semievergreen • Peninsular (coastal) Sal forest
North Eastern Coastal Plains	• Mixed deciduous forest • Mangrove of Paradeep • Peninsular (coastal) • Mangrove forest of Bhitarkanika Sal forest.
Mid-Central rugged ridge rigion of the state	• Moist peninsular valley-Sal forest • Mixed deciduous forest • Peninsular (coastal) Sal forest
Eastern-Ghats (North)	• Dry deciduous peninsular Sal forest • High level moist adjoining villages • I. tomentosa seral type of moist deciduous forest • Valley Sal forest (Moist peninsular)

* Adopted from *"State of Orissa's Environment"*, p. 35, published by Council of Professional Social Workers (CPSW), Bhubaneswar, 1994.

Table 1.2*: Districtwise distribution of forest *vis-a-vis* geographical area/population percentage of forest area/per capita forest area as on 1.4.90

District	*Land area in sq km*	*Forest area in sq km*	*% to land area*	*Population (1991 census)*	*Per capita land area*	*Per capita forest area*	*Population density*
Balasore	6311.00	258.13	4.09	279631.00	0.225	0.009	443/sq km
Bolangir	8913.00	1437.01	16.12	1703755.00	0.523	0.084	191/sq km
Cuttack	11142.00	1717.10	15.41	5503307.00	0.202	0.031	494/sq km
Dhenkanal	10827.00	5342.49	49.34	1900675.00	0.569	0.281	176/sq km
Ganjam	12531.00	7032.88	56.12	3143120.00	0.398	0.223	251/sq km
Kalahandi	11772.00	5653.59	48.02	1591984.00	0.739	0.335	135/sq km
Keonjhar	8303.00	2494.35	30.03	1315627.00	1.631	0.189	158/sq km
Koraput	26961.00	8866.79	32.90	2999903.00	0.89	0.295	111/sq km
Mayurbhanj	10418.00	5266.66	50.46	1871796.00	0.556	0.281	180/sq km
Phulbani	11119.00	5651.10	50.82	858553.00	0.295	0.658	77/sq km
Puri	10182.00	3283.27	32.24	3570192.00	0.285	0.094	351/sq km
Sambalpur	17516.00	6201.13	35.40	2688395.00	0.651	0.231	153/sq km
Sundergarh	9712.00	3989.07	41.06	1568442.00	0.619	0.254	161/sq km
STATE	15507.00	7	36.72	31512070.00	0.494	0.181	202/sq km

*Adapted from *A Decade of Forestry in Orissa 1981-90*, Office of PCCF, Orissa, p. 21.

Only on hilltops and high plateaus Montane Sub-Tropical forests occur on a very limited scale.* *Shorea robusta* Gaertn.f.(Sal) is the dominant species. The location of the state in the hemisphere is such that its vegetation constitutes both southern and northern tropical forest types. The most important feature is the merger of Teak and Sal in the forests of southern and south-eastern districts of the state. The teak belt covers five important forest divisions namely, Jeypore, Nawrangpur, Bhawanipatna, Khariar and Bolangir and Sandalwood (Chandan) in the forests of Koraput, Kalahandi and Ganjam districts. Except a few small pockets where Sal does not occur these species are found in almost all areas of the State.

Broadly the forest of the State may be divided into the following types.

THE EVERGREEN AND SEMI-EVERGREEN FORESTS

This forest type is found near sea coast and on the plateau above 700 m. in height. The important tree species met with in this area are *Shorea robusta* Gaertn.f. (Sal), *Terminalia arjuna* (Roxb. ex DC.) Wt. & Arn. (Arjun), *Mangifera indica* L. (Mango), Canes and bamboos including heavy growth of shrubs and climbers.

THE MOIST DECIDUOUS FORESTS

This type of forest occurs in areas having less rainfall and in humid conditions. The common species found in these forests are *Shorea robusta* Gaertn. f. (Sal) including associates like *Terminalia tomentosa* (Roxb. ex DC.) Wt. & Arn. (Asana), *Pterocarpus marsupium* Roxb. (Bija), *Adina cordifolia* (Roxb.) Hook.f.ex Brandis (Halanda), *Anogeissus latifolia,*(Roxb. ex DC) Wall. ex Guill & Perr. (Dhaura), *Diospyros melanoxylon,* Roxb. (Kendu), etc. Among bamboos, *Dendrocalamus strictus,* (Roxb.) Nees (Salia Baunsa) is common.

THE DRY DECIDUOUS FORESTS

This forest type is found over vast areas of the state. *Tectona grandis* Linn.f. (Teak) and *Bambusa arundinacea,*(Retz.) Willd. (Kanta baunsa) are important species seen in the area. Sal occurs here and there in

---

* (Quoted from *Geography of Orissa*, by B.N. Sinha, p. 79 published by National Book Trust, India, 3rd edition, 1999).

this belt. Other important species that are also met with in the dry deciduous forests are *Lagerstroemia parviflora* Roxb.(Sidha), *Anogeissus latifolia,*(Roxb. ex DC.) Wall. ex Guill & Perr. (Dhaura), *Acacia nilotica* (L.) Willd.ex Del (Babul), *Acacia catechu,*(L.f.) Willd. (Khaira), *Terminalia chebula* Retz. (Harida), *Sterculia urens* Roxb.(Genduli), *Terminalia bellirica* (Gaertn.) Roxb. (Bahada), *Schleichera oleosa* (Lour.) Oken (Kusum), *Madhuca indica* Gmel. (Mahua), *Bombax ceiba* Linn. (Simuli) and *Cassia siamea* Lam. (Chakunda).

LITTORAL FOREST

This forest is met with on the seashore. Orissa has a very good littoral vegetation. The common tree species grown in this belt is *Casuarina* equisetifolia L. (Jhaun).

TIDAL SWAMP FOREST

The Tidal Swamp forests are confined to the ex-Kanika and ex-Kujanga zamindari areas in Balasore and Cuttack districts respectively. The mangrove forests are confined to Dhamra, Baitarani and Mahanadi, most of the ex-zamindari forests of Kanika and Kujanga.

SIMILIPAHADA FOREST

The Similipahada forest of Mayurbhanj district covering an area of about 2,750 sq km is the most important forest of the state. It comes under the broad category of north Indian tropical moist deciduous forest type with moist peninsular Sal as the predominant species. Five different forest types are noticed in the Similipahada hills.

Tribal communities

The forest clad, undulating and hilly regions of the State of Orissa is inhabited by most of its tribal communities. These communities with 70,32,214 population constitute 22.21 per cent of the State's total population of 3,16,59,736 as per 1991 census. In respect of population strength of Scheduled Tribes, Madhya Pradesh and Orissa occupy the first and second position respectively in the country. Maharashtra comes next to Orissa. In percentage Madhya Pradesh recorded 23.26 per cent, Orissa 22.21 per cent and Maharashtra 9.27 per cent.

The distribution of the tribal population in the State is uneven. Their number is more in inland districts (old districts) like Koraput, Mayurbhanj, Sundergarh, Keonjhar and Phulbani than in coastal plains such as Balasore, Cuttack, Puri and Ganjam. The districtwise distribution of ST population as per 1991 census is given in Table 1.3.

Table 1.3: Districtwise distribution of ST population in the State (1991)

Sl.No.	*Name of the district*	*Total population*	*ST population*	*Percentage*
1.	Sambalpur	26,97,153	7,40,323	27.44
2.	Sundergarh	15,73,617	7,98,481	50.74
3.	Keonjhar	13,37,026	5,95,184	44.51
4.	Mayurbhanj	18,84580	10,90,626	57.87
5.	Balasore	28,02,417	1,97,992	7.06
6.	Cuttack	55,22,659	1,82,188	3.29
7.	Dhenkanal	19,08,907	2,32,519	12.18
8.	Phulbani	8,63,903	3,22,408	37.31
9.	Bolangir	17,07,753	3,16,797	18.55
10.	Koraput	30,12,546	16,36,079	54.30
11.	Ganjam	31.58,764	2,96,845	9.39
12.	Puri	35,90,026	1,27,347	3.54
13.	Kalahandi	16,00,385	4,95,425	30.95
	Total	**3,16,59,736**	**70,32,214**	**22.21**

Source : Census of India, 1991.

The population size of different communities of the State varies from 39 among the Chenchus to 989342 among the Kondhs as per 1981 census. There are 15 major tribal communities numbering each more than 1 lakh in population. They are the Bhuinya, Gond, Munda, Kharia, Oraon, Santal, Kolha, Bhumija and the Bathudi of northern and western Orissa, and Saora, Kondh, Sabar and Lodha, Paraja and the Bhattada of southern Orissa.

Linguistically the tribes of Orissa can broadly be divided into two categories, namely Mundari group and Dravidian group. Mundari group includes Saora, Remo, Didayi, Parenga, Juang, Karmali,

Santal, Kharia and Munda languages. Within the Dravidian family there are nine languages namely Pengo, Gondi, Kisan, Kondh, Koya, Parji, Kui, Kuvi and Kurukh or Oraon.

The tribes of Orissa are found at various levels of economic development. On the basis of primary occupation they are broadly classified into four groups such as (1) hunters and food gatherers, (2) shifting cultivators, (3) settled agriculturists, and (4) industrial workers. The first category includes the Birhor, Mankidia, Chenchu, Malhar, and Hill Kharia. These people live in the forests of Mayurbhanj, Keonjhar, Sundergarh and Dhenkanal. They lead a semi-nomadic life exclusively depending on collection of forest produce for their livelihood. They also catch small birds and animals. They live in temporary huts made out of materials collected from the forest. Their entire lifestyle revolves round the forest.

The practice of shifting cultivation is widespread in Koraput, Phulbani and Ganjam districts in southern Orissa and in the district of Keonjhar and some parts of Sundergarh in northern Orissa. The Juang and Paudi Bhuinya of northern Orissa and the Dongaria and Kuttia Kondh, Lanjia Saora, Didayi, Bondo and some Paraja, Gadaba and Parenga of southern Orissa practise shifting cultivation by bringing forest patches under cultivation usually on hill slopes. They also collect forest produce both for their household consumption and for sale.

The major bulk of the tribal population in Orissa is settled agriculturists though these people supplement their economy with activities like hunting, gathering, wage earning and agricultural labour. The Santal, Bhuinya, Gond, Munda, Ho, Oraon and the Bhumija are best examples.

A sizeable tribal population of Orissa works in mines and factories. During the past decades the process of industrial urbanisation accelerated in the tribal belts of Orissa specially in the districts of Keonjhar, Sundergarh and Koraput through the operation of mines and establishment of industries in these areas. A number of tribal people specially the Santal, Kolha, Munda and Oraon work in mines and factories mostly as unskilled workers.

Orissan tribes are generally classified into two sections, the "primitive" group and the "advanced" group. The primitive tribes are at pre-agricultural and low level of technological development, practising hunting, food gathering and shifting cultivation. The groups belonging to the first category such as hunters and food gatherers and shifting cultivators come under primitive group. The advanced group includes the settled agriculturists and industrial workers.

The primitive tribes live in comparatively inaccessible areas. They are at the lowest rung of educational development and their population size is comparatively lower than that of the advanced section of the tribal groups. Their world-view is fully in consonance with the forest ecosystem. However, for the purpose of development thirteen tribes have so far been declared by the government as primitive communities. They are Juang, Paudi Bhuinya, Bondo, Didayi, Lanjia Saora, Birhor, Lodha, Kharia, Mankidia, Kuttia Kondh and Dongaria Kondh who are segmented into different exogamous totemic clans and are mostly forest dwellers. The totems relate to plants and animals found in their forests. The social sanction prevalent in tribal communities prohibits any harm to the totemic phenomena. In other words, the social system of the tribal people provides a large measure of protection to the forest and propagation of vegetation. Clans are again divided into lineages, which function as effective corporate social units.

The family is the smallest social unit in tribal communities. It is patrilineal, that is, the descent and inheritance are through males. Residence after marriage is patrilocal. The custom of bride-price is prevalent among them. The girls enjoy considerable freedom in the matter of selection of their partners. A woman can seek dissolution of marriage on grounds of infidelity, impotence and cruelty on the part of the husband. A man can also divorce from his wife, if she is proved to be barren, disloyal to her husband, idle or diseased. However, a divorced woman can claim compensation from her husband. In household management, important decisions are taken by males. However, the women are sometimes consulted in such decisions. In day-to-day matters, both men and women take decisions jointly.

Among them the division of labour is more or less equitable. The women work side by side with the men. Work like collection of green leaf vegetables, edible fruits, roots and tubers, firewood from forest, weeding, transplantation in agricultural fields are done by women in addition to their household work and looking after the children. The men on the other hand do hard work like collection of honey, tapping of gum and resin, hunting and catching of birds and small animals in the forest, tree felling, watching the crops in the shifting cultivation land, ploughing, broadcasting of seeds in agricultural fields, etc. Wage earning is done by both men and women. In general, women share the duties and responsibilities of earning livelihood and maintaining the daily life of their families on equal terms with the menfolk. In other words, tribal women are more involved in the collection of EFP. They are the ones who are knowledgeable about the quality of fuelwoods, timber and locational details and seasonal availability of different EFP.

Symbiotic relationship of tribals with forests

Most of the tribal communities live in and around the forest. To a vast number of tribal people, the forest is their well-loved home, their livelihood and their very existence. It gives them edible fruits and leaves, honey, roots and tubers, game and fish. It provides them with materials to build their home and to practise their arts and crafts. So the forests and the tribals are inextricably interlinked with each other. Fifty per cent of the time of a tribal is spent in agricultural activities followed by collection of EFP. The income from EFP comes next to agriculture. In Orissa the average percentage of income from collection of EFP is about 12 per cent to 15 per cent. The National Commission on Agriculture estimated for the nation as a whole that employment provided by collection of EFP was 250 million man-days and by proper management it would have increased to 1000 million man-days by AD 2000.

The symbiotic relationship which exists between the tribal people and the forest does not confine to the economic sphere alone. It covers other spheres like medicine, recreational activities, social, religious and cultural aspects of their life. This relationship ensures the fulfilment of their daily needs and the protection of the environment. Since the tribal communities are almost landless or marginal farmers, they depend upon forests more for survival than on other sources of livelihood. The variety of trees and plants yielding valuable EFP and with food value are too many to identify. They are knowledgeable about such plants and the use to which they use in their daily life. They know which kind of timber is most suitable for building their houses, for agricultural implements, for furniture and for crafts and artefacts. Thatching grass, rope-making fibres, cup and plate-making leaves, trap-making reeds and bamboos, poisonous ingredients used in arrowheads and numberless such other products are collected from the forests. Their isolated existence in inaccessible areas in forests has compelled them to bank upon herbal medicine in times of illness.

In most areas the health centres are situated in far-off places, difficult of access by the tribal people. This is the reason why the tribal people still utilise herbs, the details of which are rarely disclosed to outsiders. Thus, there is no aspect of their life which has no relationship with the forest products and their habitat. Since they depend upon the forests, in totality, for their existence, there exists a balance between their own and the ecological needs. In other words, the relationship of mutual dependence ensures sustainable protection to the tribals as well as to the vegetation in the locality. This is what has been called a symbiotic relationship of mutual benefit. Since the forest serves as the very life support system of the tribals, they look upon it as theirs, and therefore, take all precautions against destruction of the forests and forest wealth. The social control, taboos and prohibition, manner of utilisation of forest products, their myths and legends reminding them of the consequences of any harm done to the nursing mother, reinforce the symbiotic relationship for their existence in peaceful harmony with nature.

2

Forest Policy, Essential Forest Produce and Tribal Development

As stated earlier, the tribal communities, particularly the primitive tribes who continue to live in the forest, depend upon the forest for their food, shelter and even to some extent clothing. The customs of tribal life including their religious practices, arts and artefacts, social fabrics and folklores have been nurtured and enriched by forests in which they are living in virtual symbiosis from the dawn of human history.

The contribution of forest to the economy of the tribal people is immense. Besides getting free all necessities of life such as fuel, fodder, wood for construction of houses and agricultural implements, medicinal herbs for curing ailments, animals and birds for hunting and roots and tubers to meet the needs of food, the tribals also earn about one-third of their income from the sale of EFP. Here comes the question of rights and privileges enjoyed by the tribal people in respect of their use of forest produce.

The subject of EFP which constitutes the theme of the study does not provide scope for a detailed analysis of evolution of forest policy starting from the Hindu and Muslim rule through British Government up to the present time of independent India. But a cursory view of the Forest Acts and Forest Policies reveals that the tribal people used to enjoy full freedom in respect to use forest products in the past. With the gradual extension of government authority over the forest there have been considerable changes in the symbiotic relationship of the tribal communities with the forest. The traditional life of the tribals are no longer recognised as rights. In 1894, these were rights and privileges, and in 1952 they became rights and concessions which later on was regarded as only concessions. For a better understanding of evolution of forest policies from the past to the present time a brief sketch of the process of development in this matter appears to be necessary. The following paragraphs give a short account of the evolution of the Forest Acts and policies.

Forest in geological ages and historical times

The paleo-botanical evidences amply testify that there were dense forests 250 million years ago. Man was evolved in the beginning of the Pleistocene age a million years ago. At this time except Rajasthan and parts of Punjab which lay buried under a swamp, all other parts of India were covered under thick forests teeming with wild life. When the Aryans came to India they cleared the forests and built shelters for themselves and for their tame animals. Being worshippers of nature they preferred their abode in forest environs and similar inspiring woodlands. It is in these settings that they were immersed in writing the Vedas and Upanishads. The epic *Ramayana* gives descriptions of dense forests in Naimisharanya, Chitrakuta, Dandakaranya and Panchavati. As time passed by, onslaught in the forest had taken place for some purpose or the other. There was a time when the area around Vrindaban was abound with such trees as Kadam and Ashok but they are conspicuous by their absence at the present time. The chronicles of Chinese pilgrims and records relating to the invasion of Alexander the Great show that there were imperial forests in the Indo-Gangetic valleys and other places. The emperors and rulers of those periods had taken steps for avenue plantations, cultivation of exotic medicinal plants and creation of exquisite gardens. The plantation of trees along the Delhi-Patna highway by Sher Shah Suri, mango groves at various places, plantation of Chinar trees in the Kashmir valley are only a few of the examples of the steps taken by the rulers in the past for the protection of forest. But for sometime the efforts of restoration of forests were reversed during the British supremacy when teak wood and certain other Indian timbers were found very suitable for building ships and when other woods were found suitable for railway sleepers needed for extending the railways from the ports to the forests in the interior. The forests were exploited for valuable forest timbers in the imperial interest. But when it was realised that the destruction of forests was detrimental to British rule in India in the long run there arose a consciousness and need to utilize the forest resources on a scientific basis.

Forest policy in retrospect

The year 1855 was the beginning of a systematic forest policy. The then Governor-General Dalhousie issued a memorandum on forest conservation. In 1856, the German botanist Dietrich Brandis was appointed as the first Inspector General of Forests to the Government of India. It was under his guidance that the forest development was organised and the first Forest Act was enacted.

FOREST ACT OF 1865

This Act was meant to regulate forest exploitation, management and preservation. For the first time, an attempt was made to regulate the collection of forest produce by forest dwellers. Basically, it was an attempt to regulate the practices of the local people under the force of law.

FOREST ACT OF 1878

This Act was aimed at increasing the control of the Government over the forests. Under this Act, the forests were divided into :

(1) Reserved forests, (2) Protected forests, and (3) Village forests, Previous claims and privileges of people over land and forest produce in the proposed reserved and protected forests were curtailed by law. The practice of tresspass and pasturing of cattle was prohibited. A duty on timber was imposed. Necessary Acts were brought into force to make it public what practices amounted to forest offences and what punishments were prescribed for committing such offences.

FOREST POLICY RESOLUTION, 1894

This Resolution declared that the sole object with which the State forests were to be administered was the public benefit. On the whole, the constitution and reservation of a forest involved the regulation of rights and restrictions of privileges enjoyed by the people in their neighbouring forest areas. The cardinal point was the emphasis laid on the advantages to be gained by the public limiting the rights and privileges of the individuals enjoyed by them as per custom and traditions.

The forests were divided into four classes :

1. Forests, the preservation of which was essential for environmental protection,

2. Forests, for the supply of valuable timbers for commercial purposes.
3. Minor forests.
4. Pasture lands.

It was categorically specified that the claims of cultivation were stronger than the claims of forest preservation and that whenever any demand for cultivable land could be supplied from forest area it should be granted without hesitation. The resolution also stressed the need for raising finances from the forests.

INDIA FOREST ACT, 1927

By decrees the Government of India gradually increased its control over the forests and the Forest Department was strengthened from time to time with a view to regulating people's right over forest lands and forest products. There was a steady increase of revenue obtained from the forest. The functions and duties of different forest officials were codified. Different forest products were listed. The items which were declared as forest produce under the Act are as follows :

Timber, charcoal, catechu, wood oil, resin, natural varnish, bark, lac, mahua seeds, myroblans, leaves, flowers and fruits, grass, creepers, reeds and moss, skins, tusk, horns, bones, silk cocoons, honey, wax, peat, surface oil, rocks and minerals.

Elaborate provisions were made to extend state control over forests. Provisions were also made for taking over the management of private forests in certain cases.

CONSTITUTIONAL CHANGES

In 1935, the British Parliament based on the Government of India Act of 1935 created Provincial Legislature but by virtue of which a dual system of government came into operation. The subject of forest was included in the provincial list under the Act. Thereafter several provinces made their own laws to regulate forests.

NATIONAL FOREST POLICY, 1952

In laying down the National Forest Policy, 1952, the Ministry of Food and Agriculture enumerated six paramount needs of the country:

1. The need for evolving a system of balanced and complimentary land use under which there would be better production and less deterioration.
2. The need for checking —

   (a) denudation in mountainous regions.

   (b) the erosion along the treeless banks of great rivers.

   (c) the invasion of seas on coastal tracts.
3. The need for establishing treelands.
4. The need for grazing opportunities, small wood and firewood.
5. The need for supply of timber and other forest produce.
6. The need for generating maximum annual revenue in perpetuity.

Under this new policy the forests were classified into :

1. Protected forests to be preserved for climatic consideration.
2. National forests to meet the needs of defence, communication and other general purposes of public importance.
3. Village forests to provide firewood, small timber and grazing.
4. Tree lands for protection of landscape.

The following statement made at the time of launching the National Forest Policy, 1952, clarifies the government views regarding the relationship of tribals with the forests

> "Village communities in the neighbourhood of a forest will naturally make greater use of its products for the satisfaction of their domestic agricultural needs. Such use, however, should in no event be permitted at the cost of national interest."

While, therefore, the needs of the local population must be met to a reasonable extent, the national interest should not be sacrificed because it is not directly discernible, nor should the right and interest of future generations be subordinated to the improvidence of the present generation. The Scheduled Areas and Scheduled Tribes Commission, 1960 appointed by the President of India under the Chairmanship of Shri U.N. Dhebar reviewed the old 1894 policy and the new 1952 policy and studied the impact of these policies on tribal communities. The findings of the Commission in respect of differences between the two policies which affected tribal communities are as follows :

1. The old policy envisages the release of forest land for cultivation subject to certain safeguards. The new policy withdraws this concession.
2. The old policy left margin for the supply of the needs of the villagers from the outlying areas in the reserved forests. The new policy decided that there should be village forests for this purpose.
3. The old policy did not touch the private forests of the tribals. The new policy applied to the same control to them.
4. The old policy did not touch free grazing in forest. The new policy sought to bring it under control.
5. The new policy admitted that persuasive and not coercive measures should be used in a sort of missionary rather than in an authoritarian manner to attempt to wean the tribals from their traditional axe-cultivation.

Forest policy creates a situation whereby the tribals who were formerly enjoying considerable freedom in respect of their use of forest product turned into subjects placed under the Forest Department. Tribal villages were no longer an essential part of the forests but were merely on sufferance. The traditional rights of the tribals were no longer recognised as rights and the symbiotic relationship with the forest was clipped for good. In 1894, the tribals were enjoying seven rights and privileges over the use of the forests, but in 1952 these became rights and concessions.

NATIONAL COMMISSION ON AGRICULTURE (N.C.A., 1976)

The Commission in its report on forestry has advocated commercialisation of forest at all costs and recommended a drastic reduction in people's rights over forests. It was stated 'Free supply of forest produces to the rural population and their rights and privileges have brought destruction to the forests and so it is necessary to reverse the process. The rural people have not contributed much towards the maintenance or regeneration of forests. Having over-exploited the resources they cannot in all fairness, expect that somebody else

will take the trouble of providing them with forest produce free of charges.' Thus, it shows that the Commission's recommendations were quite opposed to those made by the Scheduled Areas and Scheduled Tribes Commission.

CONSTITUTIONAL CHANGES

As pointed out in the previous sections the subject of forests was included in the state list as per the VIIth Schedule of the Constitution of India. But it was changed in the year 1976 whereby the subject of forest was transferred from the state list to the concurrent list through the 42nd Amendment to the Constitution of India. But by this change the powers of the State were diminished and those of the Centre were enhanced over the forests. The Government of India promulgated the Forest Conservation Ordinance in the year 1980 prohibiting the State Government from allowing the use of any forest land for any other purposes without the prior approval of the Central Government. The ordinance was listed and converted into an Act, that is, No. 69 of 1980.

THE INDIAN FOREST BILL, 1980

In accordance with recommendation of the N.C.A. a draft forest Bill was circulated in 1980 which curtailed the people's rights over forest land and produce. However, this Bill was withdrawn because public criticisms were mounted against it. A committee was appointed to reconsider the present forest policy. The committee which was set-up in 1980 emphasised the dependence of the tribal people on forests.

NATIONAL COMMITTEE ON THE DEVELOPMENT OF BACKWARD AREAS

The Planning Commission constituted a National Committee for the development of backward areas in 1980. The Committee subscribed to the views expressed by the National Commission on Agriculture and recommended curtailment of rights of tribal communities over forest lands and products.

The Committee also emphasised the importance of EFP in providing substantial income to tribal communities. It recommended removal of middle men in all business transactions and stated that EFP should not be treated as a source of revenue to the State. It was of the opinion that the economic interest can be created in the mind of the tribal communities for maintenance of forests provided substantial income accrues to them regularly from EFP. The tribal communities should be paid on the basis of the market price, not on daily wage basis, for the collection of any items of EFP. All leases for collection of EFP should be given exclusively to the co-operatives organised by the tribal communities.

THE REVIEW COMMITTEE ON THE RIGHTS AND CONCESSIONS IN THE FOREST AREAS

The Forest Department which was under the Ministry of Agriculture till 1984 constituted a committee in 1981 to review the present status of the rights and concessions obtaining in the forests administered by the government and recommended steps for effective management consistent with the scientific management of forests. The report of the committee was published in 1984. It recommended that exercise of rights and concessions should be restricted only to those tribals residing within a maximum distance of 8 km from the existing forests. It was also recommended that the beneficiaries should not be allowed to enter the forest and remove the products and that the supplies should be made through Department Depots opened outside the forest for this purpose. It also recommended that the barter and sale of forest produce obtained by the beneficiaries should be stopped.

NATIONAL FOREST POLICY, 1988

The basic objectives of the 1988 Forest Policy are oriented towards maintenance of environment stability, conservation of the natural heritage of the country, checking soil erosion and denudation of the catchment areas of the rivers, checking sand-dunes in the desert areas, increasing the forest-tree cover in the country, meeting the requirements of fuel-wood, fodder, essential forest produce and small timber of the rural and tribal population, increasing the productivity of the forests, and encouraging efficient utilisation of forest produce.

In regard to the rights and concessions the policy was very much oriented towards the development of the tribal people. It is stated that the life of the tribals and other poor people living within and near forests revolve around forests. The rights and concessions enjoyed by them shall be fully protected, their domestic requirements of fuel wood, fodder, essential forest produce and construction timber should be first charge of forest produce. These and substitute materials should be made available conveniently at reasonable prices. There is a section which gives a picture of the tribal people *vis-a-vis* forests. What is stated under the section 'Tribal People and Forest' under chapter 4 is reproduced below.

> "Having regard to the symbiotic relationship between the tribal people and forest, a primary task of all agencies responsible for forest management including the Forest Development Corporation should be to associate the tribal people closely in protection, regeneration and development of forests as well as to provide gainful employment to the people living in and around forest."

FOREST-BASED INDUSTRIES

For the first time the resolution (National Forest Policy, 1988) tries to recommend something about the forest produce consumed by the forest-based industries. It was laid down that forest-based industries should raise the raw materials preferably by establishment of direct relationship between forestry and individuals who grow it. It is suggested that the forest-based industries should provide employment to the local people and also involve them fully in raising trees and raw materials. Natural forests should not be exploited to start new industries or to expand the existing forest-based industries. It is pointed out that the practice of supply of forest produce to the industries at concessional prices should cease. Industries should be encouraged to use alternative raw materials because the supply of forest produce to industries in public and private sector has been the main factor leading to destruction and overexploitation of natural forests.

Forest administrative set-up

In Orissa, the management of forests started in the year 1883-4 with constitution of Orissa Forest Division at Angul. At that time Orissa, Bengal and Bihar were constituted as one State and Angul division started functioning under the control of Conservator of Forest of the Bengal Presidency. In 1912, Orissa was separated from Bengal and tagged up with Bihar after which two more forest divisions were created with headquarters at Puri and Sambalpur.

Orissa became a separate state in the year 1936 and four forest divisions were created namely Angul, Puri, Sambalpur East and

Sambalpur West. In that year the Forest Department was formed under a Conservator of Forests with headquarters at Angul. At that time there were 26 Princely States (Gadjats) and many zamindaris were having thick forests of their own. The ex-rulers of Princely States were very careful about the protection of flora and fauna. They used to take keen interest in the forest administration. They had qualified professional foresters to manage the forests. However, the British Government appointed a Forest Advisor to advise the ex-rulers on the proper management of forests. Dr. H.F. Mooney, one of the illustrious foresters of the time served for a long time as the Forest Advisor of Eastern States Agency under which the forest of Orissa were included. Besides the ex-rulers of Princely States there were zamindars who had excellent forest areas but these were not well managed.

After Independence the Princely States were merged with the State of Orissa as a result of which forest areas of the State increased to a large extent. Again the forest areas owned by the ex-zamindars and later on managed by the Anchala Sasan were transferred to the administrative control of the Forest Department in the year 1957. Thus, from a total area of 3,616 sq km in pre-Independence time, the forest areas increased to 67,461 sq km in 1957.

The sudden increase of forest areas necessitated changes in the administrative set-up in the State. In the year 1959 the forest administration was reconstituted with one Chief Conservator of Forests as the administrative head of the Department. The Forest Department consisted of 3 territorial circles, 1 development (functional) circle, 29 territorial divisions, 3 working plan divisions and one each for silviculture, forest utilisation and afforestation. A National Park Division was also temporarily created with the aim of developing Similipahad forests of Mayurbhanj district into a National Park.

Some of the important events in forest administration during the period from 1960 to 1990 is the formation of the Orissa Forest Corporation Ltd. for trading in timber and other forest products, creation of Similipahad Forest Development Corporation Ltd., Orissa Plantation Development Corporation Ltd., a Social Forestry Project, and the formation of Orissa Composite Board Ltd. and the implementation of a World Food Programme Project for socio-economic development through forestry activities in Orissa. Subsequently the Similipahad the Forest Development Corporation

Ltd. in the year 1990 and the Orissa Composite Board Ltd., in the year 1991 were merged with the Orissa Forest Corporation Ltd. to eliminate the overlapping of forest activities and for achieving a reduction of overhead expenditure. The organisation was renamed Orissa Forest Development Corporation Ltd.

Another important event in forestry in Orissa was the nationalisation of Kendu leaves. The item of EFP was nationalised in the year 1973 with a view to eliminating middlemen for greater revenue earning. The Forest Department and the Orissa Forest Corporation Ltd. have been jointly made responsible for the implementation of the scheme of 'wholesale nationalisation of Kendu leaves trade'. Collection and processing of Kendu leaves are done by the Forest Department with funds advanced by OFDC to the Chief Conservator of Forest (K.L.), Orissa. The OFDC handles the sale of Kendu leaves. As a result of the nationalisation of this item the State Revenue has gone up from Rs. 260.00 lakh during 1970-1 to Rs. 7435.00 lakh during 1993-4. The Kendu leaves trade has not only earned sizeable revenue to the State exchequer but also it provided employment to a large number of people who are engaged in different types of work such as bush cutting, phadi (store house) repair, collection of leaves, binding and transportation works. As the work of the Forest Department increased, a separate wing was created to handle Kendu leaf independently under the control of Additional Chief Conservator of Forest. The post was upgraded to C.C.F. in the year 1983.

At present the Forest Department functions under the Forest and Environment Department at the government level. The Principal Chief Conservator of Forests is in overall charge of forest and administration. Forest Department has four wings namely Territorial, Kendu leaves, Social Forestry and Wildlife. Each wing is managed independently by one senior I.F.S. Officer in the rank of C.C.F. under the control of P.C.C.F. There is another circle for the preparation of the working plan.

Forest policy, essential forest produce and tribal development

With regard to the symbiotic relationship between the tribal people and forest a primary task of all agencies responsible for forest management including Forest Development Corporation should be to associate tribal people closely in protection, regeneration and development of forest as well as to provide gainful employment to them. The rights of tribals over the collection of EFP may be safeguarded in such a way that the tribal people get the maximum benefit from the forest wealth.

Marketing of EFP is not yet systematically done. The LAMPS which are meant for marketing the EFP and giving the tribals a fair price are not functioning properly. Co-operatives need to be organised among the tribal people for taking up marketing of their produce by themselves. At every stage starting from collection to marketing through grading, processing, packing and transporting the goods from tribal villages to different market centres it is necessary to impart training to the tribal people and equip them with technical knowledge so that they become efficient in producing quality products with value added to the items in the interest of the tribals.

Research and development relating to various items of EFP should be undertaken by various centres of ICSIR, FRI and SFRI, etc., particularly relating to better methods of collection and grading, proper processing and utilisation of processed products and replenishment and multiplication of species.

The potential of EFP to support poverty eradication programmes is very great. Organisation of EFP Corporation taking tribal communities in the vanguard of the programme promises scope for improvement in the life of the tribal communities. It is necessary that all intermediaries like forest contractors may be substituted by institutional bodies like tribal labour co-operative societies, Sabai grass growers' co-operative societies, Tasar silkworm rearers' co-operative societies, etc. In order to make these tribal co-operative societies viable, it is necessary that different departments like Soil Conservation, Irrigation, Power including Forest Department should entrust works of their respective departments to these tribal co-operative societies so that their inner strength is built up enabling them to shoulder greater responsibility.

The Orissa forests are rich in EFP of various types. Various adverse factors such as inaccessibility, organisational inadequacy, lack of incentive, etc., stand in the way of full collection of various items of EFP. In some cases unlawful collection of certain items, inadequate storage and transport facilities and harmful tapping are responsible for scanty collection of certain EFP, sometimes leading to disappearance of EFP yielding species.

The Tribal Co-operative Marketing Development Federation (TRIFED) which has been set-up as an apex body at the national level should be given all supportive strength for its operation in respect of handling most of the tribal produce as well as all essential forest products. It has to be strengthened to function as canalising agency for export of certain EFP and also serve as a Central Nodal Agency for organising collection, processing, storage and development of oil seeds of tree and forest origin.

The National Forest Policy 1988 professes full protection to the rights and concessions enjoyed by the tribals. But it restricts these rights and concessions including grazing, and relates them to the carrying capacity of the forests. Immediate steps will have to be taken so that the rights and concessions of the tribals in forests are codified and as far as possible there should be no further infringement of these rights.

3

Collection and Procurement of Essential Forest Produce

Types of essential forest produce

FOREST resources consist of two types of products, namely (a) major forest products and (b) essential non-wood forest products. The major forest products include timber, small wood and fuel-wood and charcoal. The essential non-wood forest products include all products obtainable from the forests other than wood. Thus, EFP is composed of products of vegetable, animal and mineral origin except those minerals which are under the control of the Mining Department.

The items of essential forest products may be classified into the following categories :

1. Fibres and flosses
2. Grasses, reeds, cane and fodders
3. Essential or volatile oils
4. Tannins and dyes
5. Gums, resins and oleoresins
6. Drugs and medicines
7. Edible products
8. Oil seeds
9. Leaves
10. Animal, mineral and miscellaneous products.

Annexure I (a) gives different items of EFP available in Orissa under different categories including other information.

FIBRES AND FLOSSES

The fibres may be classified into three categories namely (a) soft fibres like jute and flax which are derived from bast of stems, (b) hard fibres collected from leaves of Manila hemps and Sisal, and (c) surface fibres collected from the surface of stems, leaves and seeds like cotton. The fibres are used in textiles including netting and cordage. These commercial fibres are of the types of cotton, jute flax, ramie and hemp. There are plating and weaving fibres. These fibres are used for making baskets, straw hats, mats, chair seats, and other forms of wicker works. Certain fibres are used as brush fibres, filling fibres and paper-making fibres and in the manufacture of natural fabrics. Generally these fibres are hard in texture.

The trees which give fibres from stems are many, but those which are available in Orissa forests are *Calotropis gigantea* R. Br. (Arakha), *Acacia leucophloea* Willd. (Gohira), *Bauhinia vahlii* W.&A. (Siali), *Butea monosperma* (Lam.) Taub. (Palas) giving fibres from young roots, *Ficus benghalensis* Linn. (Bara) giving fibres from aerial roots, *Ficus religiosa* Linn. (Aswatha), *Sterculia villosa* Roxb. (Kodala), *Sterculia urens* Roxb. (Genduli) and *Kydia calycina* Roxb. (Kapasia).

Fibres from leaves

Plant leaves yield fibres. Some of the species whose leaves yield fibre are as follows :

Agave cantula Roxb. (Baramasi) yielding fibres from the leaves which are extracted by the process of retting, *Agave sisalana* Perr. ex Engelm. (sisal fibre), *Caryota urens* Linn. (Salap) known as Indian Sagopalm are leaf fibre yielding plants.

Flosses

Flosses are used for a wide variety of purpose which include manufacture of life belts and as a stuffing in mattresses, pillows and cushions. The most important floss yielding species found in Orissa forest is *Bombax ceiba* Linn. (Semul). Capsules of these trees yield a floss which is soft and strong, widely used for life saving appliances at sea and also as a stuffing in mattresses and pillows. *Calotropis gigantea* R. Br. (Arakha) is also another species of shrub type, the fruits of which give fine, silky, soft, glossy and resilient floss materials.

GRASSES, REEDS, CANES AND FODDERS

The most important grass of great commercial value is *Eulaliopsis binata* (Retz.) Hubbard (Sabai grass). This grass is extremely suited for manufacture of paper and pulp. It is also used for making ropes with which carpets are made. Another type of grass of commercial value is *Thysanolaena maxima* (Roxb.) Kuntz (Phuljhadu or hill brooms). This species is used in broom making. Fodder grasses are of numerous types. Some grasses are exotic while others are native to different parts of India. Most of them grow wild and some of them are cultivated. The main fodder grasses are as follows :

1. *Andropogon ascinodis* Clarke. : It is a gregarious species growing in the hills of north-east India.
2. *Andropogon pumilus* Roxb. : It is a gregarious annual grass. This species is a good fodder and preferred by cattle both in green condition and as hay.
3. *Bothriochloa ischaemum* : This is a fairly good fodder species found everywhere in the countryside.
4. *Chrysopogon aciculatus* (Retz.) Trin. : It is found in the plains and low hills and on village pasture lands.
5. *Brachiaria mutica* (Forssk.) Stapf.: It is used for silage, hay and grazing.
6. *Cenchrus ciliaris, Linn.*
7. *Dichanthium annulatus*
8. *Panicum antidotale,Retz.*

Fodder Trees

There are many trees the leaves of which are relished by cattle and goats. Some of the common fodder yielding trees are *Acacia catechu* (L.f.), Willd. (Khaira), *Acacia nilotica* (L.) Willd. ex Delile (Babul), *Azadirachta indica* A. Juss. (Neem), *Bombax ceiba* Linn. (Semul), *Butea monosperma* (Lam.) Taub. (Palas), *Dendrocalamus strictus* (Roxb.) Nees. (Baunsa), *Ficus benghalensis* Linn. (Bara), *Ficus religiosa* Linn. (Aswatha), *Gmelina arborea* Roxb. (Gambhari), *Lagerstroemia parviflora* Roxb. (Sidha) and *Polyalthia longifolia* (Sonn.) Thw. (Debadaru).

Grasses used for matting

There are a number of grasses which can be split and used for weaving mats. These include

(i) *Cyperus compressus* Linn.

(ii) *Saccharum munja* Roxb. (Sarakanda)

(iii) *Typha angustata* Bory & Chaub.

Grasses used for making ropes

The most important grass used for making rope is *Eulaliopsis binata* (Sabai) which has been already referred to above. The other types of grasses used for this purpose are *Saccharum munja* Roxb. (Sarakanda) and *Themeda arundinacea* (Roxb.) Ridely.

Grasses of various types may be used for other purposes such as the manufacture of chairs, stools, baskets and screens, for *khus-khus* screens of aromatic scented mats and for housebrooms.

Canes

Over 350 cane species are found in India. They are the stems of climbing plants of the Genus Calamus and other related genera such as *Ciratolobus korthalsia*. The canes are used for a wide variety of purposes such as making of ropes and cables, for making furnitures, walking sticks, umbrella handles, and for making baskets. Canes are split into fine strips which are used for making seats, mats, baskets and other fine wicker works. Some of the important cane species found in Orissa are *Calamus guruba* Buch.-Ham. *Calamus rotang* Linn. and *Calamus viminalis* Willd.

ESSENTIAL OR VOLATILE OILS

Essential oils are volatile odoriferous liquids occurring in many plants. They evaporate when exposed to air. Different parts of the plant such as fruits, flowers, leaves, bark, root, wood and seed yield such essential oils of aromatic smell. It is found out that over sixty families of Indian plants yield essential oils. Various methods are adopted for the extraction of essential oils from plants. Some of these methods are distillation, expression and extraction by solvent using enfleurage (extracted under ordinary temperature) and maceration (extracted from flowers such as rose).

Some of the essential oils are as follows:

1. *Grass oils* : The important grass oils include (a) citronella oil, (b) ginger grass oil, (c) lemon grass oil, (d) palmarosa oil, and (e) Vertiver oil.
2. *Wood oils* : The important wood oils are (a) agar oil, (b) cedar oil, (c) pine oil, and (d) sandalwood oil.
3. *Leaf oils* : The important leaf oils are (a) camphor oil, (b) cinnamon leaf and bark oil (c) eucalyptus oil, and (d) mint oil.
4. *Root oils* : The important root oils are (a) costus oil, (b) *khus-khus* oil.
5. *Flower oils* : The important flower oils are (a) champa (*Michelia champaca* Linn.) (b) ketaki, and (c) golab (rose) oils.

TANNINS AND DYES

Tannins are secretions of plant tissues. They are used mainly to render the hides and skins resistant to decomposition and make them strong and flexible. The bark of many trees yield bark tans. Standing trees are not usually stripped off their bark to extract the tannins. Generally the barks are obtained from felled trees. The important species which yield bark tannins are as follows :

1. *Acacia nilotica* (L.) Willd. ex Delile (Babul): The bark of these trees yield tannins of superior quality.
2. *Anogeissus latifolia* (Roxb. ex DC.) Wall. ex Guill & Perr. (Dhaura) : The leaves of this species are used for extraction of tannins.
3. *Cassia fistula* Linn. (Sunari) : The bark of this species gives tannin.
4. *Ceriops discandra* (Griff) Ding Hou: Both bark and leaves of this species yield tannins.
5. *Shorea robusta* Gaertn.f (Sal) and *Terminalia Arjuna* W.&A. (Arjun): Barks of these trees produce tannin of reddish brown colour of good quality.
6. *Fruit tans* : The fruit of many plant species give tannins which are used locally for commercial purposes. Some of

the species giving tannin yielding fruits are as follows.

(i) *Terminalia chebula* Retz. (Harida) : By far the most important of the myrobolan is the fruit of *Terminalia chebula* Retz. Its fruits yield tannin. The best time for collection of fruits from the point of view of tannin content is January.

(ii) *Terminalia bellirica* (Gaertn.) Roxb. (Bahada) : Its fruits yield tannin.

(iii) *Emblica officinalis* Gaertn. (Amla) : This plant species yield tannin from its fruits, leaves, and twig and stem barks.

7. *Leaf tans* : Leaf tans are generally used by local tanners. The species which yield leaf tans are *Anogeissus latifolia* (Roxb. ex Dc.) Wall. ex Guill & Perr. (Dhaura) and *Carissa spinarum* L. (Khirkoli). Other well-known leaf tans are collected from *Emblica officinalis* Gaertn., (Amla) *Garuga pinnata* Roxb. and Terminalia species.

Dyes

Many plant tissues secrete vegetable dyes. There are wood dyes, bark dyes, flower and fruit dyes, root dyes and leaf dyes. The plants of the Artocarpus family such as *Artocarpus heterophyllus* Lam (Panas) and *Artocarpus lacucha* Roxb. ex Buch. -Ham. yield wood dyes. The wood is powdered which is boiled in water and extracted. By this process a bright yellow dye is obtained. The bark dyes are obtained from Acacia species and Terminalia species. The flowers and fruits of many plant species yield valuable dyes which are more important than the dyes obtained from wood, bark and root. The species which comes under *Butea monosperma* (Lam.) Taub. (Palas) yields a bright yellow colour dye from its flowers. Similarly *Crocus sativus* Linn. is the source of yellow colour dye collected from its flowers. The fruits of *Mallotus philippinensis* (Lam.) Muell.-Arg. (Kamalagundi) yield a red dye which is the most popular dye of India. It is used for dyeing silk and other fabrics. The root dyes are obtained from certain plant species such as *Morinda pubescens* Sm. (Achu), *Punica granatum* Linn. (Bedana). The leaves of some plant species also yield dyes. Chlorophyll which is present in most plants is being used widely in the manufacture of foodstuff, soap, mouth fresheners and dental cream. One of the well-known species yielding leaf dyes is *Lawsonia inermis* Roxb. (Mehendi). Its leaves yield an orange dye Henna.

Gums, resins and oleoresins

Gums and resins are exuded by plants mainly from the stem and also from roots and leaves. Those products are exuded as a normal process and also as a result of injury or disease to the bark or wood of the plant.

Gums are a sticky substance. They dissolve or are softened in water. They are known as true gums. Resin has a plastic consistency, soluble in alcohol but unaffected by water. These are known as hard resins. Oleoresins are mixed with a fair percentage of essential oils. The gum resins are a mixture of gums and resins. There are also gum oleoresins which are composed of oleoresins with a small quantity of gum.

Gums

Gums are formed as a result of disintegration of plant tissue through a process called Gummosis. Natural gums emanate from plant as liquid which on exposure dries up into translucent amorphous flakes. These having the least colour and highest adhesive power are most valuable. These varieties are used for clarification of liquors, finishing of silk and preparation of quality water colours. Intermediate grades are used in confectionery, cosmetics, pharmaceuticals and printing inks and in dyeing. The cheaper grades are used in adhesives and in printing industries. Two kinds of gums are noteworthy for commercial purposes — Gum Arabic and Gum Tragacanth. The Gum Arabic is imported mostly from Sudan. It is collected from *Acacia senegal* Willd. The Babulgum collected from *Acacia nilotica* (Linn) Willd. ex Del. can serve as a substitute for real Gum Arabic. The gum collected from *Anogeissus latifolia* (Roxb. ex DC.) Wall. ex Guill & Perr. (Dhaura) which is known as Ghattigum can also be used as a substitute of real Gum Arabic. Gum Arabic which is imported from Sudan (about Rs. 1000 per ton) is used for dyeing and block printing on fabrics.

Genduli gum obtained from *Sterculia urens*, Roxb. species or Gum Karaya, as it is called, is used as a substitute for the real Gum Tragacanth. The state which produces the largest quantity of Genduli gum (about 3500 tonnes) is Madhya Pradesh. The technical know-how for processing the gum is lacking in India. Therefore, whatever quantity of gum is exported is all unprocessed and the countries getting the Indian gum get it processed to make it suitable for use and also for exporting to other countries.

Further, due to defective methods of collection, cleaning, grading and packing, the produce does not fetch a good price. Research and extension in these aspects and training to the collectors can add value to the produce. Indiscriminate tapping and clandestine cutting and felling have resulted in diminution of the gum yielding plants. Replenishment and plantation programmes are conspicuous by their absence.

The gums are exuded in the following manner :

1. Spontaneous exudation in dry seasons.
2. Charred trees affected by forest fire exude gum.
3. Artificial incursion allows gum to flow out.
4. Heavy and intensive tapping causes gum to exude.

Many plants exude gum. Of them the trees account for a large portion. Some important species are (a) *Acacia catechu* (L.f.) Willd. (Khaira), (b) *Acacia modesta* Wall. (c) *Acacia senegal* Willd. (not available in Orissa), (d) *Anogeissus latifolia* (Roxb. ex DC.) Wall. ex Guill & Perr. (Dhaura), and (e) *Bauhinia retusa* Roxb. There are various other gum yielding species of secondary importance such as (a) *Albizia chinensis* (Osbeck) Merr. (b) *Anacardium occidentale* Linn. (Kaju), (c) *Mangifera indica* L. (Mango), and (d) *Spondias pinnata* (L.f.) Kurz.

Resins

Resin secretes from cavities or passages in many plant species. When the resins exude from the bark they get hardened on exposure to air and sunlight.

Resins are used for a wide variety of purposes. Certain types of resins are melted and mixed with seed oil or turpentine oil to form oil varnish. Some resins are dissolved in alcohol to form spirit varnish. They are dissolved in alkali to form soaps. They are also used in medicine and pharmacy, for sizing papers, for incense and in the preparation of sealing wax.

The resins can be classified as Hard resins and Oleoresins.

Hard Resins

Some of the commercial hard resins are : Copal, Dammar, Amber, Lacquer, Sandarac and Mastic.

Oleoresins

Oleoresins contain considerable amount of essential oil together with resin material. Some of the commercial oleoresins are Balsam, Turpentine, Copaiba and Elemi.

Gum resins

The gum resins are a mixture of both gum and resins. These gum resins are of great commercial value. Some types of gum resins are used for adding flavour to curries and sauces. One variety of gum resins is used as incense. Some others are used in medicines.

Some of the leading tree species of yielding hard resins, oleoresins and gum resins leading tree species of are mentioned below:

1. *Boswellia serrata* Roxb. ex Colebr : Its trunk exudes gum oleoresin. It is used mainly as an incense and as medicine to get relief from rheumatism and nervous diseases.
2. *Canarium strictum* Roxb. : Its bark exudes resin. Resin flows from vertical incisions for six months yearly and this flow goes on for about 10 years. It is mainly used in the manufacturing of varnishes, bottling wax, etc.
3. *Shorea robusta* Gaertn.f : The Sal tree yields an oleoresin. The tree is blazed and the resin exudes out. Each mature tree can yield on an average 5 kg of resin every year. Resin also exudes naturally from injuries caused by Sal heartwood borers.

Drugs and medicines

Man from the earliest time has been using medicinal herbs to get relief from various diseases. Roots, tubers, stems, barks, leaves, fruits and flowers of various medicinal plants are used as curative of different diseases. The important EFP of medicinal importance which are collected from the forests of Orissa are Amla, Baghanakha, Bahada,

Baidanka, Banahaldi, Banakulthi, Banatulsi, Chakunda, Char seeds, Chireita, Dhatuki flower, Gaba, Harida, Indra jaba, Kaincha, Kurchi Kochila (*Strychnos nux-vomica* Linn.), Landa beguli, Lodhra, Mankadakendu, Mahu (honey), Nageswar flower, Palash, Pita alu, Putrani R.S. root, Sikakai.

EDIBLE PRODUCTS

Hundreds of edible forest products in the form of roots and tubers, fruits and flowers, pulps and seeds are available in different forest areas. These are known to the people and the tribal people in particular who collect these edible products. Some of these wild edible products are eaten raw and some cooked at home. Due to the shortage of land in the tribal areas the crops grown in the lands are in short supply and therefore they depend upon hunting and food gathering from the forests in and around of which they live. The starchy rhizomes, roots and tubers, green leaves are generally cooked and eaten. There are different types of fruit trees like Amba (mango), Panasa (jackfruit), Kendu, Bel, Kaitha, Tala (palm), Sitafala, Tentuli (tamarind) Jamun, Barkoli, Khajuri (datepalm) and other varieties of palms which are eaten either in the raw form or when ripe. Seeds of various fruits like kaju, mango, jackfruit, sal, char, etc., are relished very much by the tribal people. Saps tapped from many tree species like Salap (Sagopalm) and Khajuri (datepalm) quench their thirst for intoxicating drinks. Mahua flower is brewed to get a liquor which they are extremely fond of. Eggs of birds, larvae of insects particularly tasar (silkworm) and white ants are a delicacy of tribal people. Honey is a favourite food and they have acquired great expertise in its collection. These edible wild products are available to the tribal people round the year, particularly in the lean months starting from April up to the end of October.

OIL SEEDS

Oil seeds are some of the most important items of forest produce. There are about more than 100 oilseeds bearing trees in the country and many of them are found in the Orissan forests. Oil gives energy. Energy consumption in India is on the increase. The annual consumption of 649 million tonnes of oil equivalent in 1973 has increased to 2019 million tonnes of oil equivalent in 1992. Orissa is in short supply of its oil requirement. At the rate of 40 gram per head daily it requires 2143 million tonnes of oil for the entire state population against the production of 0.949 million tonnes. Some of the important oil yielding species are as follows :

1. Arakha (*Calotropis gigantea* R. Br.)
2. Agara (*Argemone mexicana* Linn.)
3. Ashwagandha (*Withania somnifera* (L.) Dunal)
4. Bhalia (*Semecarpus anacardium* Linn.)
5. Cashew nut (*Anacardium occidentale* Linn.)
6. Jatropha (*Jatropha curcas* Wall.)
7. Karanja (*Pongamia glabra* Vent.)
8. Kusum (*Schleichera oleosa* Lour. Oken.)
9. Mahua (*Madhuca indica* Gmel.)
10. Mango kernel (*Mangifera indica* L.)
11. Neem (*Azadirachta indica* A. Juss.)
12. Oil Palm (*Elaeis guinensis* Jacq.)
13. Polang (*Calophyllum inophyllum* Linn.)
14. Sal (*Shorea robusta* Gaertn. f.)
15. Sirisa (*Albizia procera* (Roxb.) Benth.)

In addition to these species there are plants, the seeds of which are used for medicinal purposes. These are Anata, Gilo, Banatulsi, Banakulthi, Charseeds, Chakunda, Babul, Morda, Baidanka seeds, Shikakai seeds, Palash and Indrajaba seeds. The oil extracted from these items are used for medicinal purposes as well as for making soaps. (See Appendix for details)

The oils can be classified into three groups namely, non-drying oils, semi-drying oils and drying oils. The oil extracted from castor seeds and ground nuts (which is a SAP) belong to the category of non-drying oil. The oil extracted from sesamum and cotton seeds fall into the category of semi-drying oil. The drying oils are those which are extracted from lin seeds.

Leaves

Like other EFP, leaves of many trees and plants occupy a very important position in the life of human beings. These are leaves which are used for medicinal purposes, leaves as edible items and leaves from which dyes are extracted. Leaves of certain tree species like *Shorea robusta* Gaertn.f. (Sal) and palm are used as thatching materials to cover the roofs of houses. The leaves of Siali creeper, Sal and Kendu trees are of great economic value. Leaf utensils are manufactured from Sal and Siali leaves. Kendu leaves are used for making bidis (leaf cigarettes). These items are in great demand in the rural areas and provide employment to thousands of people. The leaves of datepalm make good mats and this occupation keeps many women engaged in their own houses.

ANIMAL, MINERAL AND MISCELLANEOUS PRODUCTS

A number of animal products are obtained from forests. Some of such products collected from the forest of Orissa include lac, cocoon, honey and wax.

Lac

The tiny lac insects lacca secrete lac, primarily as protective covering. It has a parasitic existence living on different species. Some of the host plants of the insect are *Acacia nilotica* (L.) Willd. ex Delile (Babul), *Albizia lebbeck* (L.) Benth., (Sirisa) *Butea monosperma* (Lam.) Taub. (Palash), *Cajanus cajan* (L.) Huth. (Kandula), *Garuga pinnata* Roxb., *Schleichera oleosa* (Lour.) Oken (Kusum) and *Ziziphus mauritiana* Lam. (Barakoli). Only Kusumi lac is collected from the Orissa forests.

Silk

Tasar cocoons are collected and cultivated in Orissa. The *Antheraea mylitta D* (tasar silkworm) feeds on leaves of *Terminalia arjuna* (Roxb. ex DC.) Wt. & Arn. (Arjuna), *Shorea robusta* Gaertn. f. (Sal) and *Terminalia tomentosa* (Roxb. ex DC.) Wt. & Arn. (Asana). The tribal people in different parts of Orissa are engaged in tasar silkworm rearing from times immemorial. The women, in particular, play a greater role in this activity.

Honey and wax

Bees collect nectar from many different flowers in the forests and store it as honey. Mainly two species of bees produce honey in the wild. They are rock bees (*Apis dorsata*) and Indian bees (*Apis indica*). A single comb may yield 2 to 4 kg of honey and round about 1 kg of wax. The hill Kharias of Similipal hills of Mayurbhanj district are expert honey collectors. They are skilled in reaching the beehives wherever they are in the high hills.

Seasonality of collection

The tribal people irrespective of age and sex are engaged in collection of forest products both for their own consumption and for sale. The edible wild products like fruits, roots, tubers and green leafy vegetables are mainly collected as food items whereas other products such as seeds, honey, arrowroot, tamarind, hill broom, etc., are sold or are exchanged for their daily necessities like cereals, salt, kerosene, clothing and ornaments. Forest collection, though not a main source of income for a majority of tribal people specially among the settled agriculturists and industrial workers or shifting cultivators, is a major supplementary source of income and provides scope for engagement almost round the year, particularly during the lean period starting from April to the end of October.

Items like mango, sal seeds, ripened fruits, oilseeds, etc., are collected by women and lac, honey, bark, gum, etc., by men. Collection of honey is a difficult job and the honey collectors have to climb the hills with the help of ropes to reach the beehives. Therefore it is exclusively done by men.

Table 3.1 shows the availability of different types of EFP in different seasons. Very few items are available in the months of July to September after which the number of items increases and the largest number of items is available from January to May, the period conforming to the period of plenty and festivals. The slack season begins from July till October when the growth of plants takes place and it is forbidden to interfere in the process by tapping the plants. This period coincides with the period of scarcity of food.

Table 3.1: Availability of different types of EFP in different seasons.

Sl. No.	*Essential forest produce*	*Availability by Months*											
		Jan	*Feb*	*Mar*	*Apr*	*May*	*June*	*July*	*Aug*	*Sept*	*Oct*	*Nov*	*Dec*
1.	Seeded Tamarind	*	*	*	*	*							
2.	Deseeded Tamarind	*	*	*	*	*							
3.	Hill broom	*	*	*	*								*
4.	Myrobalan (Harida)	*	*	*								*	*
5.	Honey		*	*	*	*	*	*	*				
6.	Siali leves	*	*										*
7.	Nux Vomica	*	*										*
8.	Marking Nut	*	*	*								*	*
9.	Cleaning Nut	*	*								*	*	*
10.	Antudi Flower				*	*							
11.	Dhatuki Flower	*	*	*									*
12.	Swan Climber	*	*	*	*	*							*
13.	Thorn Broom	*	*	*									*
14.	Genduli Gum	*	*	*	*								
15.	Dhaura Gum	*	*	*	*								
16.	Amta Gum	*	*	*	*								
17.	Bahada Gum	*	*	*	*								
18.	Other Gum	*	*	*	*								
19.	Char Gum	*	*	*	*								
20.	Lac	*	*			*	*	*				*	*
21.	Nageswar flower	*	*										
22.	Sunari Bark				*	*							

Sl. No.	*Essential forest produce*	*Availability by Months*											
		Jan	*Feb*	*Mar*	*Apr*	*May*	*June*	*July*	*Aug*	*Sept*	*Oct*	*Nov*	*Dec*
23.	R.S. Root				*								
24.	Siali Fibre				*	*							
25.	Mushrooms (Edible)							*	*	*	*		
26.	Bhalia Koli				*	*	*						
27.	Fenfena bark			*	*	*							
28.	Simuli Cotton			*	*	*							
29.	Bhuin neem									*	*	*	*
30.	Khilua creeper				*								
31.	Bahada	*	*	*								*	*
32.	Amla	*	*	*								*	*
33.	Broom grass	*								*	*	*	*
34.	Sal seeds					*	*						
35.	Soap nut					*	*						
36.	Tamarind seed				*	*							
37.	Thatch grass	*	*										
38.	Cane				*	*							
39.	Mahua seed				*	*							
40.	Karanj seed		*	*									
41.	Kusum seed					*	*						
42.	Char seed				*	*	*						
43.	Babul bark					*	*				*	*	
44.	Resin (Sal)			*	*	*	*						
45.	Patal Garuda	*									*	*	*
46.	Shikakayi		*	*									
47.	Fenfena fruit									*	*		
48.	Ashok bark	*	*										*
49.	Mushroom (Poison)									*			

Sl. No.	Essential forest produce	Availability by Months											
		Jan	Feb	Mar	Apr	May	June	July	Aug	Sept	Oct	Nov	Dec
50.	Kendu leave				*	*							
51.	Arrowroot		*	*	*								
52.	Arjuna bark	*	*										*
53.	Sal leaves	*	*								*	*	*
54.	Mango kernel				*	*	*	*					
55.	Mahua flower		*	*									
56.	Pita Alu										*	*	*
57.	Chireita	*	*										
58.	Mankada kendu fruit				*	*							
59.	Rohini flower		*	*	*								
60.	Bee wax				*	*	*						

Methods of processing and practice of replenishment

Very little processing of EFP is done at the local level. The products are sold raw as soon as these are collected. Naturally little efforts are being made for value addition. Whatever is described below are all theoretical and not practised by any tribal or local people.

FIBRES

The method of extraction differs from species to species. The process by which fibres are loosened from the woody plant tissues is called retting. The plant stems are kept under water for fermentation to take place. The gummy substance is removed by applying water manually. Then the stems are broken and the fibres scrapped. Underretting keeps some gummy substance attached to the stem whereas overretting weakens the fibres. Not all species are required to be kept immersed under water, because in some cases the fibres become useless due to the application of water. Fibres from the leaf are separated by the same process of retting. The fibres which are extracted from the leaves are thoroughly washed and then dried in the sun. Grasses are collected manually and dried and then arranged into bundles.

TANNINS

Wood tan

Wood tan is basically obtained from barks. No standing tree is stripped off its bark to extract tannins. The rainy season is avoided for harvesting. The stripped-off bark is dried in the sun for some days and put in a wooden vat to which steam is applied for extraction of tan liquor from the bark. The liquor is purified and mixed with other substances to give the desired quality and colour to the leather. In Orissa such species as *Cassia fistula* Linn. (Sunari), *Acacia nilotica* (L.) Willd. ex Delile (Babul), and *Acacia catechu* (L.f.) Willd. (Khaira) supply barks containing wood tannin. Tannin is also extracted from leaves. These are dried and powdered from which tan substances are extracted.

Kamala gundi

Ripe fruits are collected in gunny bags and beaten and shaken till the red powder comes out. This process is applied to *Mallotus philippensis* (Lam.) Muell. -Arg. (Kamala gundi). Sometimes fruits are soaked in water and then dried and then the dye is extracted.

GUMS, RESINS AND OLEORESINS

Gums

Gums are obtained in various ways. Some of the methods are mentioned below.

In some cases the gums exude spontaneously in hot summer months and in other dry seasons. Gums exude from those trees which are charred by forest fire. The gums are collected by hand. Large lumps are broken into pieces. Barks, stems, leaves, sands, etc., are removed by hand as far as possible. Depending upon the quality of the gum, some grading is done manually.

Gum is collected by artificial incisions. In Orissa gum is extracted from *Acacia nilotica* (L.) Willd. ex Delile (Babul), *Acacia catechu* (L.f.) Willd. (Khaira), *Anogeissus latifolia* (Roxb. ex DC.) Wall. ex Guill & Perr. (Dhaura) and *Sterculia urens* Roxb. (Genduli), *Terminalia bellirica* (Gaertn.) Roxb. (Bahada) and Buchanania lanzan Spr. (Char).

Acacia catechu (L.f.) Willd. (Khaira) does not give any gum during the rainy season. The gum begins to ooze out freely during the dry season, usually from the injured parts of the tree. These trees can be tapped for a couple of years.

Anogeissus latifolia (Roxb. ex DC.) Wall. ex Guill & Perr. (Dhaura) yields gum which is called Ghatti. The gum oozes out naturally from the wounds during the summer months.

Bauhinia retusa Roxb. (Kanchana) gives gum of superior quality. A deep blaze is made with the help of a sharp knife. The gum oozes out from the blaze during the hot season. A tree is tapped for four to five years and then given a rest.

Sterculia urens Roxb. (Genduli) is the source of Gum Karaya. The gum which exudes from the tree is of four grades. The finest grade is either white or colourless, the second grade is either pinkish or dirty white. The third grade is black in colour. The fourth grade is dust like. The main properties of gum karaya are : It is not soluble in water, but swells into a soft mass. It is a substitute for gum tragacanth.

Resins

Resins of different types are collected from different species. *Boswellia serrata* Roxb. ex Colebr. (Salai) gives Salai gum. A piece of bark is peeled off at a certain height of a fully grown tree exposing ducts containing resin. The first collection is made after two weeks. Salai gum exudes from the blaze in long tears. It hardens after about four days and retains its golden colour and transparency. The tapping is done in November and goes on up to July until the monsoon sets in.

Shorea robusta Gaertn.f. (Sal) is a source of oleoresin. Resin exudes from a blaze on the tree trunk and solidifies on exposure. Blazing is done at the higher level to exude more resin from September to November.

EDIBLE PRODUCTS

Hundreds of edible wild plants occur in the forests. Different parts of different plants are edible. Some edible items are eaten raw, some are baked and some other are cooked. Many fruits are eaten when they are unripe and some other fruits are eaten ripe. Many types of edible roots and tubers are collected by digging them out from the underground with the help of digging sticks or sharp-pointed iron crowbars. Leaves of many plants are plucked and cooked. Pulp of fruits is eaten without cooking them. Kernels like mango kernels are dried, grinded into powder from which cakes are prepared by baking.

The tribal people have acquired knowledge regarding seasonality of various items of wild edible products. For example, when the leaves of many tuberous plants turn yellow it is ready for digging. Arrowroot which is very common wild edible product is dug out from the ground when its leaves turn yellow.

Harida, Bahada, Amla, Tamarind, etc., are collected using a uniform method. For collection more than one person is required. Generally a family consisting of three to five members is employed in the job. Gunny bags, baskets, broomsticks, bamboos are required for collection. Sometimes some people climb up and shake the branches so that the fruits fall on the ground. These fruits are heaped in a particular place with the help of broomsticks and thereafter brought home in gunny bags.

The seeds and roots of *Bauhinia vahlii* W.&A. (Siali) are eaten. The seeds are eaten raw or fried. The roots are boiled and eaten as a substitute for potatoes. The fully-grown plant yields edible parts. Roots are dug out by crowbars and the seeds are collected when the fruits are ripe. Some plant species contain delicious sap which is tapped by incision. *Caryota urens* Linn. (Salap) is a sago palm which gives a refreshing juice collected from the flower stock which ferments into toddy or palmwine. Palms having an age of more than 15 years are tapped and a palm tree of full growth yields about 15 litres of juice in 24 hours. Dateplam also yields juice which is drunken fresh or after it is fermented. The juice is boiled to yield jaggery. A good quality sago is obtained from the pith of the stem. The sago is dried and pulverised to make cakes out of it. *Dendrocalamus strictus* (Roxb.) Nees (a variety of bamboo) yields seeds which fall on the ground. These are eaten either raw or after making a gruel out of seed flour. The young shoots of the male fruit containing sweet pulp is relished by the tribals. These fruits are either plucked or collected from the ground. *Ficus glomerata* Roxb. (Dimiri) fruits are eaten vegetables when they are ripe. The fruits are plucked by climbing up the tree and collecting them in bags or baskets.

Oil seeds

The oil seeds which are collected from the forests are Kusum, Neem,

Karanja, Mahua, Banatulsi, Narda, Nageswar, Gillo, Banahaldi, Chakunda, Babul, Polanga, Mango Kernel and Sal seeds. Mainly two methods are used for extraction of oil — (1) Expression and (2) Extraction. The method of expression involves the following steps.

The seeds are cleaned and all impurities are removed. The kernels are separated from the outer shells either by crushing or running a heavy roller over the seeds. Oil is extracted by cold pressing or by hot pressing. In the latter case the crushed seeds are boiled and kept in two containers which are placed one upon the other, face to face. These containers are placed on a wooden plank kept horizontally at a height above the ground and with another plank one end of which is fixed to a tree trunk, the containers are pressed by lowering the free end of the upper plank to touch the end of the lower plank.

Besides this traditional method of extraction, oil presses (*ghani*) are used to extract oil. The oil press is operated by harnessing bullocks to rotate the beam which presses the oil seeds in the Ghani.

EXTRACTION BY SOLVENT METHOD

In Orissa there are four solvent extraction units which are attached to OFDC and TDCC. The oil seeds collected by these public sector undertakings are extracted at these units. The names of these oil extraction units are Orissa Oil Industries, located at Sasana and Rairangpur. Utkal Oil located at Ambaguda and Priti Oil Ltd. at Rengali.

Leaves

Previously it has been mentioned that leaves of many different species are used as edibles. Besides, there are leaves from which dyes are extracted. There are also leaves of different plants which are of medicinal importance. In addition to these uses of the leaves, there are leaves of great commercial importance. Kendu leaves, for example, for which Orissa has a name in the country are used for rolling bidis. The State Government gets millions of rupees of royalties out of the sale of Kendu leaves. Similarly Sal and Siali leaves are of great economic value. These products fetch millions of rupees to the State exchequer. All these leaves are plucked manually by thousands of labourers employed for this purpose. These are dried and kept in bundles in stores maintained by lease holders.

ANIMAL PRODUCTS

Mainly three items are worth mentioning in this section. These are Lac, Honey and Wax, and Cocoons.

Lac

The study on EFP shows that only Kusumi lac is collected by the TDCC. There are two crops of Kusumi lac, one crop called 'Aghani' and the other is called 'Jethwi'. The Kusumi crops are mostly grown from the tree called *Schleichera oleosa* (Lour.) Oken (Kusum) and with Kusum brood lac on a few host plants. Some of the host plants are *Acacia nilotica* (L.) Willd. ex Delile, and *Cajanus cajan* (L.) Huth. In Kusumi strain inoculation for Aghani crop is done in June and July and harvesting in January and February, and for Jethwi crop inoculation is done in January and February and harvesting in June and July. The Jethwi supplies brood for the major Kusumi crop. Since the period of development of the Jethwi crop falls entirely into the summer months, it very often undergoes failure because of the extreme heat.

The figures for the years 1993-4 the quantity collected was 209.04 qtl. whereas it has increased to 280.00 qtl. in the year 1994-5. There was a time when the State was producing 1381 MT half a century ago (1947-8). Almost two decades ago Orissa was producing on an average 300 MT of Kusumi lac every year. The present average production as stated by the study group on the lac sponsored by the NABARD is about 80 MT. But the present study shows that the procurement of lac was at the level of 25 MT.

It shows that the production of Kusumi lac in Orissa suffered a great setback. The reasons for the decline in production are many. There is non-availability of brood-lac. Although two brood farms are maintained in the State, the quantity of brood lac production is insignificant. Out of nine lac growing districts of the state the TDCC has taken the districts of Keonjhar, Sundergarh and Mayurbhanj for development of lac cultivation.

Besides the shortage of brood lac, adverse climatic conditions and price fluctuation, dominance of intermediaries, continuous traditional method of cultivation, inadequate extension services and restrained availability of credit all taken together have been working adversely against the lac cultivation.

Honey

Many tribal people collect honey mainly for sale. The Hill Kharia of Mayurbhanj district are expert bee-wax and honey collectors. In a dark night the Kharias climb trees or hills where beehives are traced out. The hives of hills are usually found in the precipitous side of the hill which cannot be reached on foot. A ladder fixed to a tree at the top of the hill is lowered up to the hive. A man comes down with the help of this ladder up to the hive and moves a torch emitting smoke around the hive to scare away the bees. He pokes at the comb with a stick and breaks it so that honey and pieces of hive fall into a basket which is held under the hive. One hive yields about four to six litres of honey and one to two kg of wax.

Honey is usually collected from January to April, and wax is collected as by-product. After the honey is squeezed out, the hives are boiled in water. This mixture is poured into a gunny bag placed on a pit dug in front of the house. This strained material is poured into a pan placed over a pot of boiling water. By this process the water content is steamed out and the wax deposited as a sediment. Thus, the wax is made ready for sale.

Tasar Cocoons

Tasar cocoons are collected from the forest and also these are cultivated near the village sites. The tasar silkworm moth of Orissa *Antheraea mylitta*.D. produces cocoons which are the source of the silk are locally called Bobei and Nadia. The insect is a denizen of the forest inhabited by tribal communities. The tasar silkworm feeds on Asan, Arjun and Sal.

The seed cocoons are collected from the forests during the months of June and July. They are sold during August and September to the rearers as seeds. The moths emerge during the period from August to October and the layings are collected and retained for rearing. The newly-hatched larvae of August and September spin cocoons in the branches of host plants by the end of October or early November. Two crops can be obtained in a year. Sometimes three crops are possible, if the moths are bivoltine or trivoltine in character.

It is a practice that forests are leased out by the Forest Department in the tasar tract to the rearers in the interest of the industry. A moderate fee is collected from the rearers as Dalkati (cutting of the twigs). The Adapahis (a tasar-growing plot having hundreds of host plants) are under cultivation since times immemorial and the Adapahi of one person is not trespassed by any other person. The trees are topped each year to get fresh leaves for the rearing during the next year.

A rearer temporarily shifts his residence to his Adapahi from his home. He leads a life of a hermit following self-imposed religious restrictions and obligations as per custom and tradition. The rearer and his whole family sit under trees with bows and arrows to scare away the birds and other enemies who eat up the moths.

The silk worms crawl out to the leaves of the trees in which the packets containing hatched eggs are placed. The worms grow larger and larger, and spin cocoons in the tree. From the colour of the cocoons the termination of spinning by the silkworms is indicated. After 8 to 10 days the cocoons are harvested and sold out in the weekly market or to the Tasar growers co-operative societies. In Orissa generally the cocoons are steamed so that the silkworms inside the cocoons are killed. These steam-stifled cocoons are sun-dried and later stored in bags for sale.

An overview of collection practices

It is observed that the tribal people who are closely linked with forest and whose life pivots round forests, are employed as day labourers for which they are paid daily wages not in term of market price of the produce which they collect. Except the nationalised items private trading is in vogue in respect of most of the items. Even in the case of nationalised items the real growers or rearers who are mostly tribals do not turn up to compete with other bidders and ultimately the non-tribal merchants succeed in the auction sale and enjoy the profits and the tribals are expropriated from such benefits.

Massive amount of money is realised from different agencies by royalties payable to the government. A portion of it according to the quantum of areawise collection of EFP should go to the Tribal Sub-Plan for utilisation in development of tribal people. The main objective of setting up co-operative bodies was to eliminate private trading and middlemen and involve tribal people in the business for their gain. But it is hardly fulfilled. Various agencies which have been given lease or entrusted with the task of collecting EFP engage, at places, local private traders in the work of collection and buy the produce from them as well as sell them to the same traders, thereby the tribals were largely eliminated instead of the private traders.

Very little processing of any EFP is undertaken by the private collectors to add value to the EFP in the interest of the tribal people. In fact, the entire marketing and whatever little processing is done is taken care of by the wholesalers. The study shows that preliminary processing of most of the items of EFP can be done by the tribals at their doorsteps, because no complicated process is involved at the primary stage of processing.

The whole analysis boils down to the fact that the procuring agencies have not tried to involve the tribal communities in the preliminary processing due to shortage of manpower and financial resources needed for this purpose. Of all the negative factors the will to take up the exercise at the tribal level is the most important. The foremost concern of agencies is to achieve bulk procurement and make payment of revenue to the State exchequer. In this situation, charged with anxiety to obtain revenue for the Forest Department the collecting agents are made to deliver the commodities to purchasing agents and the payment obtained from the latter is passed on to the Forest Department, thus, leaving no scope for processing. In other words, there is no plan at any level to have an infrastructure for tribal development throughout the process of collection and marketing transactions relating to EFP.

There is a great need for training and guidance to the tribals in respect of the methods of collection and processing of items in the simplest way such as the elementary steps of tapping, blazing, drying, grading and packing, etc. Institutes like the Tribal Research Institute and the Social Forestry Training and Research Institute could be entrusted with the task of organising training courses for the tribal communities and other collecting agencies.

4

Procuring and Marketing Agencies

THE tribal people do not get a fair price for the EFP which they collect from the forests. The local merchants and middlemen exploit them in price as well as in measure and weight. In order to give a fair price to the tribal people for their SAP (Surplus Agricultural Produce) and EFP, the government has set-up some marketing organisations in the tribal areas to buy the SAP and EFP from them and supply their daily necessities at a fair price, so that they are made free from the clutches of the exploiters.

Marketing organisations

The marketing organisations which deal with EFP and carry on business in these items are (i) The Orissa Forest Development Corporation Ltd. (OFDC), (ii) Tribal Development Co-operative Corporation Ltd. (TDCC), (iii) Utkal Forest Products Limited (UFPL), (iv) Large Size Adibasi Multipurpose Co-operative Societies (LAMPS), (v) Tribal Co-operative Marketing Federation of India Limited (TRIFED), Orissa Rural Development and Marketing Society (ORMAS), and the (vii) Tikabali Agency Marketing Co-operative Society (TAMCS). Besides these co-operative organisations, there are several agencies which are connected with the business of EFP either in groups or in individual capacity.

Government policy on lease of EFP

In order to eliminate middlemen from the business of EFP and protect the tribal people from the exploitation of these middlemen, the practice of private lease in all items of EFP was prohibited. In practice, certain other items of EFP have been given on lease to private persons. The small- and medium-scale industries depending on EFP items for their raw materials were made to have business linkage with the above-mentioned co-operative agencies for supply of raw materials. In pursuance of the above-mentioned policy decisions of the Government, the following arrangements have been made since December 1990 in regard to the settlement of lease of EFP items in the state.

1. The Tribal Development Co-operative Corporation Ltd. (TDCC) will have the exclusive right to operate the lease of 4 EFP items namely Tamarind, hill broom, Mahua flower (now discontinued) and Honey in all the 27 forest divisions of the State.
2. M/s. Utkal Forest Products Ltd., a joint sector company under the Industrial Promotion and Investment Corporation Ltd. (IPICOL) will operate the lease of 39 identified EFP items in all the forest divisions of the state.
3. TDCC will operate the lease of all the EFP items excluding the 29 EFP items earmarked for UFPL, in 19 forest divisions namely Khariar, Nawarangpur, Jeypore, Kalahandi, Rayagada, Nayagarh, Ghumsar North, Ghumsar South, Parlakhemundi, Baripada, Karanjia, Keonjhar, Sundergarh, Deogarh, Bamra, Bonai, Bolangir, Sambalpur and Rairakhol.
4. In the remaining divisions, namely Puri, Dhenkanal, Angul, Athagarh and Athamalik, the Orissa Forest Development Corporation Ltd. (OFDC) will operate the lease of EFP items except the 29 items for TDCC Ltd. in Phulbani, Baliguda and Boudh forest divisions.
5. Tikabali AMCS will operate the lease of EFP items except the 29 items identified for the UFPL and 4 items for TDCC Ltd. in Phulbani, Baliguda and Boudh forest divisions.

However some minor changes have been made subsequently. The details of allotment of areas and items of EFP are given in Table 4.1.

Table 4.1: List of procuring agencies/lessee of MFP(EFP) along with area of operation and items of MFP(EFP) settled with

Sl. No.	*Agency/Lessee*	*Area of Operation (Forest Division)*	*Items of MFP (EFP)*	*Remarks*
1.	The Orissa Forest Development Corporation Ltd. Bhubaneswar.	Nawarangpur, Jeypore, Bolangir, Khariar, Parlakhemundi, Ghumsar North, Ghumsar South, Rayagada, Puri, Athagarh, Athamalik, Angul, Dhenkanal, Deogarh, Sambalpur, Bonai, Karanjia, Keonjhar, Baripada, Sundergarh, Rairakhol, Chilika WL, Chandaka WL, Satkosia WL, Badaroma WL, Sunabeda WL.	Sal Seed	
		Athagarh, Dhenkanal, Angul, Athamalik, Puri, Karanjia, Keonjhar, Baripada.	Nux Vomica, Siali leaf, Hill broom Resin, K.creeper, Cane, Myrobolan, Honey, Arrowroot, Sal resin.	
2.	Tribal Development Co-operative Corporation of Orissa Ltd.	Nayagarh, Phulbani, Boudh, Bamara, Baliguda, Kalahandi.	Sal seed	
		All 27 Forest Divisions except Phulbani, Baliguda, Boudh F.D.	Tarmarind	
		All 27 Forest Divisions except Koraput and Lamtaput Range of Rayagada F.D.	Hill brooms	
		All 27 Forest Divisions except Keonjhar, Baripada, Karanjia Forest Divisions &	Honey	

Sl. No.	*Agency/Lessee*	*Area of Operation (Forest Division)*	*Items of MFP (EFP)*	*Remarks*
		Ghumsar North. Khariar, Nawarangpur, Jeypore, Kalahandi, Rayagada, Nayagarh, Ghumsar North, Ghumsar South, Parlakhemundi, Baripada, Karanjia, Keonjhar, Sundergarh, Deogarh, Bonai, Bamara, Bolangir, Sambalpur, Rairakhol.		
3.	Utkal Forest Products Ltd.	All 27 Forest Divisions such as Nayagarh, Ghumsar South, Ghumsar North, Jeypore, Nawarangpur, Bolangir, Kalahandi, Khariar, Rayagada, Parlakhemundi, Sundergarh, Sambalpur, Bamara, Bonai, Deogarh, Rairkhol, Keonjhar, Karanjia, Baripada, Phulbani, Baliguda, Boudh, Angul, Athamalik, Athagarh, Dhenkanal, Puri.	Kusum seed, Karanj seed, Gaba seed, Nageswar seed, Gilo seed & cover Banakulthi, Landa Baguli, Baghanakhi, Anata, Polang, Neem seed, Palash seed, Siali seed, Benachar, Chakunda, Baidanka, Babul seed, Basil, Makhan seed, Shikakai seed, Mango kernel, Madha seed, Indrajaba, Banahaldi, Banatulsi, Kamalagundi, Chiranji dana, Jute seed.	All items excluding 29 items identified for M/s. Utkal Forest Products Ltd.

Sl. No.	*Agency/Lessee*	*Area of Operation (Forest Division)*	*Items of MFP (EFP)*	*Remarks*
4.	Agency Marketing Co-operative Society Tikabali.	Phulbani, Baliguda & Boudh Forest Divisions	Tamarind, Marking nut, Mat reeds, Arrowroot Lac, Honey & Wax. Cleaning nuts, Genduli gum, Harida, Bahada & Amla, Babul & other gum sabai grass, Thatch grass Simuli cotton, cane, Nux Vomica,Siali fibre.	Three years from 1.10.95 vide letter 20856 dt. 11.10.95.
5.	Gunpur LAMP	Gunpur Range of Rayagada F.D.	Nux Vomica, Tamarind, Marking and Clearning nut, Honey, Wax, Dhataki flower, Siali leaves, Simuli cotton, Soap nut, Mersinga leaves	
6.	Mujagada Forest Produce Marketing Co-operative Society.	Ghumsar North F.D. Suruda and Buguda Range of Ghumsar South Forest Division.	Siali leves and Plates, Sabai grass	
7.	Kunduli LAMP	Koraput and Lamtaput, Range of Rayagada Forest Division.	Tamarind, Myrobalan, Marking & Cleaning nut, Hill brooms, Putrani root.	Till 1996-7
8.	State Tasar & Silk Co-operative Society.	Angul, Bolangir, Rairakhol, Bonai, Bamara, Keonjhar, Dhenkanal Sadar, Baripada, Karanjia,	Tasar cocoons	3 years from 1.10.93 to 30.9.96.

Sl. No.	*Agency/Lessee*	*Area of Operation (Forest Division)*	*Items of MFP (EFP)*	*Remarks*
		Athagarh, Khariar Athamalik, Boudh, Sundergarh, Deogarh, Dhenkanal (Kamakhya-nagar).		
9.	Sawaramal Gadodia Rajgangpur.	Bonai, Boudh, Phulbani, Baliguda, G. North, G. South, Kalahandi, Sunder-garh.	Lodhamedha & Phenphena Bark	5 years from 1.10.91 to 30.9.96.
10.	Sawaramal Gadodia Rajgangpur.	Deogarh, Parla-khemundi, Nayagarh, Dhenkanal & Rayagada.	- do -	Till 30.6.96
11.	Sawaramal Gadodia Rajgangpur	Baripada & Karanjia.	- do -	- do -
12.	Narasingha Agarwala Bagdia.	Jeypore, Angul, Athamalik, Keonjhar.		- do -
13.	Debashis Ray, Sambalpur.	Sambalpur, Bamara, Rairakhol.	- do -	- do -
14.	Sawaramal Gadodia, Rajgangpur.	Rairakhol, Deogarh, Rayagada & Kalahandi.	Dantarilata	2 years from 1.1.94 to 31.12.95
15.	Suman Enterprises Bhubaneswar.	Athagarh, Nawarang-pur, Bonai, Sundergarh.	Lodhamedha, Phenphena, Dantarilata.	5 years from 1.6.95 to 31.5.2000
16.	Satyanarayana Parida Balasore.	Balasore & Portion of Baripada.	Eksira fruit, China Alu, Rohini fruit, Mushroom Sidha fruit, Atandi fruit, Dry grass.	3 years from 1.10.94 to 30.9.97.

Sl. No.	*Agency/Lessee*	*Area of Operation (Forest Division)*	*Items of MFP (EFP)*	*Remarks*
17.	Biren Soren Chandua	Baripada & Karanjia	Eksira fruit, China Alu, Rohini fruit, Mushroom Sidha fruit, Atandi fruit, Dry grass.	5 years from 1.10.94 to 30.9.99
18.	Bhagwan Das Dodka	Bolangir	Lodhamedha & Phenphena bark.	2 years from 1.8.95 to 31.7.97.
19.	Govinda Kumar Agarwal	Khariar	Lodhamedha & Phenphena bark	2 years from 1.8.95 to 31.7.97.
20.	M/s Dokichina Gurukul & Son & Co. Rayagada.	Raygada, Narayana Patna, Kashipur Ranges of Rayagada F.D.	Tamarind	5 years from 1.10.95 to 30.9.2000.
21.	Abdul Razak Khan	Athagarh	Cane	2 years from 1994-5 to 1996-7.
22.	S. Goswami Bangiriposi	Portion of Baripada & Karanjia F.D.	Lodhamedha & Kurchi	Till 30.6.96

Procuring agencies and organisational set-up

Orissa Forest Development Corporation Ltd. (OFDC)

Prior to 1962 the harvesting of forest resources was done through private contractors. This system was not satisfactory because it encouraged illicit fellings and removal of forest resources, exploitation of the workers (mostly of tribal communities) and default on payment of royalties. Taking into consideration these points, the government decided to set-up a fully government owned and managed

Corporation for the proper management and development of forests in the State.

The Orissa Forest Corporation Limited, the first public sector undertaking in the forestry sector of the country was incorporated in September, 1962 as a Public Limited Company under the Companies Act 1956 with an authorised capital of Rs. 2 crore. It started functioning from October 1962 by taking up trade in timber, firewood and other products with a paid-up capital of Rs. 25 lakh. The Corporation took up the trading of Kendu leaves in 1965-6 and Sal seeds in 1973-4. When the Kendu leaf trade and Sal seed collection was nationalised in the year 1973 and 1983 respectively, the Corporation assumed the charge as the sole selling agent of Kendu leaves and was entrusted with the collection of Sal seeds from a major part of the state. With the expansion of its business, the authorised capital was raised to Rs. 5 crore in the year 1981-2. The paid-up share capital which was entirely subscribed by the State Government, was raised to Rs. 128 lakh by 1984-5. By the year 1987-8, the Corporation was in a position to pay annually about Rs. 48 crore to the exchequer as royalty out of its total turnover of Rs. 80 crore.

OBJECTIVES

The main objectives for which the Corporation was established are as follows :

(i) To undertake scientific management and harvesting of forest resources and to carry on business in various forest produce so as to generate maximum financial resources for the State.

(ii) To plant, grow, cultivate, propagate, produce and raise plantations of all kinds including commercial and horticultural crops and to look after the marketing of the produce.

(iii) To encourage and establish forest-based industries.

(iv) To generate more employment, particularly in the rural sector and elevate the socio-economic condition of the rural population, mostly tribals.

Two other corporations, namely Similipahada Forest Development Corporation Ltd., and Orissa Plantation Development Corporation Ltd., were merged with the Corporation in the year 1990 with an authorised capital, of Rs. 2 crore and Rs. 20 crore respectively. The paid-up capital, however, of the said Corporations as on 1.10.1990 was Rs. 0.55 crore and 1.46 crore respectively.

With the merger of the Corporation the name of the Orissa Forest Corporation was changed to Orissa Forest Development Corporation Ltd. (OFDC) with effect from 14.11.1990. Later, in the year 1991, the Orissa Composite Board Ltd. was also merged with OFDC.

ORGANISATION

The OFDC is administered by a Board of Directors headed by a full-time Chairman who is an officer in the rank of Additional Development Commissioner. The Managing Director next to the Chairman in order is a senior officer of the Indian Forest Service in the rank of Chief Conservator of Forests. There are three Directors of which one is in charge of commercial activities, one for plantation and one looks after the finances of the Corporation. Each Director is assisted by subordinate officers designated as Marketing Managers. There are Chief Audit Officer, Drawing and Disbursing Officer and Accounts Officer under the control of the Finance Director. Other officers of this rank are of the category of Vigilance and Security Officer, Officer on Special duty and Chief Personal Officer and Adviser.

The Commercial Directorate consisting of timber, EFP and industrial activities is managed by five commercial zones located at Baripada, Bhubaneswar, Sambalpur, Bolangir and Berhampur. Each zone is under the charge of a Chief Executive in the rank of Conservator of Forests. There are 31 Commercial Divisions and 3 outside sale centres located at Kolkata, Delhi and Chennai. The Plantation Directorate of the Corporation is managed under three zones located at Bhubaneswar, Sambalpur and Koraput. In all there are 17 plantation divisions under these zones.

ACTIVITIES

The OFDC deals with Kendu leaves, Sal seeds and a few other EFP such as Nux Vomica, Siali leaves, honey, resin, arrowroot, Harida, Bahada, Amla, hill broom, K. creeper and Sabai grass. Kendu leaves are nationalised and therefore the CCF (KL) through the Forest Department collects the leaves. In the case of Sal seeds which is another monopoly item, both OFDC and TDCC collect the Sal seeds from their respective allotted areas and take care of marketing of the item. As regards the small EFP items the specified areas from which these are collected by the OFDC are Puri, Athagarh, Dhenkanal, Athamalik and Angul. The lease of honey, arrowroot and Sal resin was awarded in favour of OFDC in 1991 in three Forest Divisions namely Karanjia, Keonjhar and Baripada.

The Kendu leaf collection is a major EFP item running to on an average annual collection of 5,00,000 qtl. of leaves. The State gets a revenue of Rs. 4,800 lakh every year. Likewise Sal seeds are a major item of EFP. More than 24,00 MT of Sal seeds are collected by OFDC alone. It is a kind of item of EFP which gives a bumper crop in some years and much less yield in some other years. In some years the royalty paid to the government becomes as low as 62 to 70 lakh and in other years it shoots up to round about 300 to 400 lakh. As regards to other small EFP items, OFDC has been collecting these items for the last five to six years. The total quantity has never gone above 100 qtl. giving a sale value not exceeding Rs. 1,50,000. This is in exclusion of broomsticks and canes which fetch not more than same amount Rs. 1,50,000. The details of economics of Kendu leaf and Sal seeds trade is given in project profile. Annexures II, III(A) (B) (C) and IV show the year-wise procurement, disposal, sale value, etc. of Kendu leaves, Sal seeds and other EFP by OFDC respectively.

The working of Kendu leaf as stated earlier is managed by Forest Department (Kendu-Leaf Wing) and OFDC, the latter handling the sale and arranging the entire finance for the collection whereas the former is procuring the Kendu leaves. Annexure V shows the quantity of collection of Kendu leaves and expenditure on different heads such as bush cutting, drying, storage, etc. from the year 1984 to 1993. The divisionwise procurement of Kendu leaves from the year 1990 to 1994 is given in Annexure VI.

COLLECTION OF MAHUA FLOWERS

Prior to the nationalisation of Mahua flower the trade on this item was being regulated under the Orissa Excise (Mahua Flower) Rules, 1976, framed under the Orissa Excise Act, 1915. Initially, the Forest, Fisheries and Animal Husbandry Department entrusted the TDCC with the exclusive right of collection of Mahua flower throughout the State. Thereafter the government settled the lease for the collection of Mahua flowers in favour of TDCC and OFDC in 14 Forest Divisions and 13 Forest Divisions respectively. Subsequently, the Forest and Environment Department revised the allotment of Forest Divisions in the year 1991, for the collection of Mahua flowers in favour of TDCC and OFDC as the following :

TDCC	*OFDC*
1. Nawarangpur	1. Dhenkanal
2. Jeypore	2. Angul
3. Rayagada	3. Athamalik
4. Phulbani	4. Puri
5. Baliguda	5. Nayagarh
6. Boudh	6. Sambalpur
7 Boudh	7. Sambalpur
8. Baripada	8. Bamara
9. Karanjia	9. Rairakhol
10. Sundergarh	10. Deogarh
11. Keonjhar	11. Parlakhemundi
	12. Ghumsar North
	13. Ghumsar South
	14. Bolangir
	15. Khariar
	16. Kalahandi
	17. Athagarh

Subsequently the the Forest and Environment Department nationalised the Mahua flower in the year 1991 and OFDC and TDCC were appointed as collecting agents for the purchase and sale of Mahua flowers. The government fixed the purchase price of Mahua flowers @ Rs. 3.00 per kg. At that time the prevailing purchase price was Re. 1.00 per kg. However, the cost value of this item including collection price, operation cost, godown rent and establishment cost came more or less to Rs. 5.00 per kg. The OFDC and TDCC faced many problems in selling the Mahua flowers. At that time the rate of Mahua flowers in Ranchi area bordering the procurement area in Orissa was about Re. 1.00 per kg. The middlemen in that area purchased Mahua flowers from their locality and sold them in Orissa through tribal agents. Therefore, a huge amount of Mahua flowers was purchased by the TDCC and the OFDC.

But they could not get the buyers for such a huge stock as the rate was high compared to other neighbouring States. OFDC and TDCC were forced to make distressed sale at a lower cost varying from Re. 0.50 to Re. 1.00 per kg. A sizeable quantity was dumped by the OFDC to avoid godown rent. As a result the OFDC and TDCC sustained a loss of about Rs. 20 crore and Rs. 9 crore respectively. These organisations expressed their inability to trade on this item in future.

During 1992 the Forest Department has passed order that Mahua flowers shall cease to be specified as forest produce and their procurement and trading should be regulated under the existing Excise Law with effect from the next procurement season that is March 1992 and traders will be given the licence by the Excise Department for the collection and trading of Mahua flowers as per previous practice.

Tribal Development Co-operative Corporation of Orissa Ltd. (TDCC)

The TDCC, registered under the Orissa Co-operative Corporation Act, 1962, has been functioning since 1973 as an apex co-operative organisation for the development of the tribal communities of the State. The objectives of the TDCC are as follows :

1. To procure SAP and EFP collected by the tribals and pay them a remunerative price and arrange for marketing.
2. To supply essential commodities and other consumer goods to the tribal people at a fair price.
3. To arrange for the processing of the procured commodities, to add value to the items with a view to increasing profitability, and providing employment to the tribal people.
4. To advance consumption-cum-production credit to the tribal people for raising crops and also for meeting their consumption needs.

MANAGEMENT

The management of the Corporation is vested with a Managing Committee. The State Government nominates the members for three years. The President of the TDCC functions as the President of the Managing Committee. Besides, the Principal Secretaries of the Forest and Environment Department and Finance Department are members of the Managing Committee as nominees of the government. The Commissioner-cum-Secretary to Government, Welfare Department, Co-operative Department, Food Supply and Consumer Welfare Department, Chief Conservator of Forests, Registrar of Co-operatives, Director of SC and ST, a representative of the State Co-operative Bank, Presidents of Tikabali Agency Marketing Co-operative Society, and Gobindapalli Forest Marketing Co-operative Society, MLA, Gunpur and a woman representative are also members of the Managing Committee. The Managing Director of TDCC functions as the member secretary.

The Corporation has at present 240 LAMPS and other primary societies, 47 Panchayat Samitis and State Government, and 3 individuals as members. The Corporation comprises 3 divisions, 15 branches and 130 permanent procurement centres. The head office of the Corporation is located at Bhubaneswar. The structural network is as follows :

	Division	*Branch offices*
1.	Keonjhar	(i) Udala, (ii) Keonjhar, (iii) Lahunipara, (iv) Sundergarh.
2.	Berhampur	(i) Nayagarh, (ii) Baliguda, (iii) Parla-khemundi, (iv) R. Udayagiri
3.	Sunabeda	(i) Sunabeda, (ii) Rayagada, (iii) Jeypore, (iv) Muniguda, (v) Kashipur, (vi) Umerkote, and (vii) Bhawanipatna

The Corporation has 130 procurement centres, 74 fair price shops and 66 godowns located at different places in the State. The godowns have a total storage capacity of 35, 998 MT. The godowns are constructed by TDCC and the cost of construction is borne by different organisations such as (i) the National Co-operative Development Corporation (NCDC) - 50 per cent as loan, (ii) Co-operative Department - 25 per cent as share capital and 20 per cent as subsidy, and (iii) TDCC - 5 per cent.

The godowns are under the control of branch managers posted at branch offices.

EQUITY BASE

The authorised share capital of the Corporation is Rs. 20 crore against which the paid-up share capital as on 31.8.95 is 13.56 crore as indicated below.

Table 4.2: Paid-up share capital of TDCC

(i)	State Govt. in Welfare Department	Rs. 11.85 crore
(ii)	Govt. of India, Ministry of Welfare	Rs. 0.50 crore
(iii)	State Govt. in A & C Department	Rs. 0.49 crore
(iv)	Co-operative Societies	Rs. 0.02 crore
(v)	Panchayat Samitis	Rs. 0.30 crore
(vi)	State Govt. in Welfare Department participation in PS FS scheme	Rs. 0.40 crore
	Total	**Rs. 13.56 crore**

Owing to continuous loss sustained by the Corporation, its share capital has been completely eroded and the business activities at present are being carried on by borrowing. The borrowing by the Corporation as on 1.10.1994 stands as the following:

Table 4.3: Loan particulars of TDCC

	Loan particulars	*Rs. in lakh*
1.	Loan from Welfare Department towards margin money	64.86
2.	Loan from Welfare Department for augmentation of collection of tree borne oil seeds	123.23
3.	Loan from F & CS Department	76.48
4.	Ways and Means advanced by government	325.15
5.	Loan from NCDC through OSCB & RCS (O) construction of storage godowns and purchase of transport vehicles.	53.60
6.	DRI loan from Andhra Bank	93.75
7.	Loan from OK & VI Board	75.24
8.	Loan from SBI	
	(a) On cash credit	1051.50
	(b) Against fixed deposit	7.50
	Grand Total	**1871.31**

FINANCIAL ASSISTANCE RECEIVED

Financial assistance received by TDCC in the form of share capital investment, grants-in-aid, subsidy and managerial assistance from the

State Government from the years 1972-3 to 1995-6 (up to 31.8.95) is given below.:

Table 4.4: Financial assistance received by the TDCC from the year 1971-2 to 1995-6 (Rs. in lakhs)

Year	*Share capital*	*Grants-in-aid*	*Subsidy*	*Managerial assistance*
1971-2 Taken over from Forest Produce Marketing Co-operative Society (T.D.)	2.04			
1972-3	11.81	—	—	3.01
1973-4	3.21	—	6.30	13.92
1974-5	25.00	6.00	8.15	16.00
1975-6	—	30.00	2.82	16.00
1976-7	83.07	—	1.00	21.33
1977-8	31.75	7.15	3.23	11.07
1978-9	88.96	—	2.61	16.00
1979-80	103.77	—	4.26	—
1980-1	15.03	—	8.38	—
1981-2	53.00	—	2.40	1.77
1982-3	51.99	—	1.592	—
1983-4	50.00	—	1.07	—
1984-5	56.00	—	5.134	1.45
1985-6	87.855	—	6.32	—
1986-7	78.00	12.48	—	—
1987-8	12.267	10.00	9.358	—
1988-9	56.00	29.00	—	—
1989-90	56.00	10.00	—	—
1990-1	60.00	28.00	2.134	10.00
1991-2	67.68	29.35	—	10.00
1992-3	63.00	30.70	—	10.00
1993-4	85.00	33.82	—	20.00
1994-5	21.25	49.28	—	20.00
1995-6 (Up to 31.8.1995)	71.75	—	—	7.00

BUSINESS TURNOVER

Principal items of procurement among the forest produce are Tamarind, Sal seeds, myrobolan, Sal and Siali leaves and hill brooms, etc., which constitute the bulk of procurement. Itemwise procurement of EFP by TDCC is given during last 10 years in Annexure VII and the procurement price rate is given in Annexure VIII(A). Annexure VIII(B) shows the quantity and value of EFP procured by the TDCC in the years from 1992-3 to 1994-5. Districtwise procurement of EFP by TDCC of Orissa during 1990-1 may be seen at Annexure IX.

For the procurement of the above items the TDCC opens temporary procurement centres besides fair price shops which also function as procurement centres. In all about 800 temporary procurement centres are opened to cover the remote areas for collecting EFP every year. LAMPS are also appointed as the agents for the procurement of EFP.

The TDCC has been assisting LAMPS each year in procurement of EFP. The yearwise involvement of LAMPS in the procurement of SAP/EFP is as follows:

Table 4.5: No. of LAMPS Procuring EFP for TDCC during 1985-6 to 1993-4

Year	*No. of lamps*	*Procurement value (Rs. in lakhs)*
1985-6	33	22.18
1986-7	29	21.58
1987-8	63	52.34
1988-9	54	81.00
1989-90	22	19.28
1990-1	31	62.68
1991-2	50	96.71
1992-3	36	25.55
1993-4	8	36.87

The LAMPS are being given a commission of 2 per cent in respect of EFP. More LAMPS are not able to procure EFP on behalf of TDCC on account of their past default in payment of outstanding advances against them.

The TDCC has been endeavouring to pay higher procurement rate for EFP. The business turnover of the Corporation during the period from 1984-5 to 1994-5 is given in the following table:

Table 4.6: Business turnover of TDCC from 1984-5 to 1994-5

Year	*Procurement*		*Consumer*	*Total*
	SAP	*MFP(EFP)*	*business*	*(Rs. in lakh)*
1984-5	171.97	189.16	168.87	530.00
1985-6	204.26	176.83	237.38	618.47
1986-7	151.88	120.81	200.65	473.34
1987-8	256.42	135.51	190.78	582.68
1988-9	134.02	417.88	185.33	737.23
1989-90	163.39	64.28	133.83	361.50
1990-1	513.08	361.23	187.18	1061.49
1991-2	55.50	1013.99	145.26	1214.75
1992-3	429.68	221.00	141.18	791.86
1993-4	90.69	321.83	187.51	599.26
1994-5	38.18	330.52	276.85	645.55

PROFIT AND LOSS

The TDCC is running in loss since its inception. The year-wise position of profit and loss of the Corporation as per final audited account figures from the year 1989-90 to 1993-4 is as follows.

Table 4.7: Yearwise position of profit and loss

(Rs. in lakh)

Year	*Grass profit/loss trade account (Rs.)*	*Net profit/loss (Rs)*
1989-90	(—) 27.17	(—) 153.21
1990-1	(—) 126.71	(—) 74.60
1991-2	(—) 645.92	(—) 964.81
1992-3	(—) 39.15	(—) 506.81
1993-4	(—) 81.12	(—) 340.25

The above statement would show that in spite of earning profit by trading generally the TDCC is sustaining net loss due to various reasons beyond the control of the management of the Corporation. The major reasons of loss are as follows:

1. *Heavy interest burden on borrowing*: Own fund of the Corporation having been totally depleted, the TDCC has to depend

mainly upon loans from commercial banks for the procurement of SAP/EFP items. The interest on loan for working capital annually comes to about Rs. 200 lakh.

2. High overheads: At present TDCC has a staff strength of about 495 out of a total strength of 619. The Corporation expends about Rs. 1.81 crore annually on staff. As against this, the government has been releasing a sum of Rs. 20 lakh per annum towards managerial subsidy which is not sufficient.

3. Heavy royalty: The TDCC has been paying Rs. 12.13 lakh per annum from 1982-3 towards royalty based on the royalty calculation for the year 1981-2 with 10 per cent increase per annum on EFP items except Sal seeds. Total royalty on EFP including Sal seeds comes to the tune of Rs. 40 lakh to Rs. 50 lakh per year. This amount has gone up to Rs. 78 lakh in view of fixation of revised royalty by the State Government by making 10 per cent increase every year to the royalty amount of 1981-2. However, after consideration of the request of the Welfare Department, TDCC, and the Forest and Environment Department have decided that complete exemption from royalty is not possible. However the royalty may be realised from the TDCC in respect of various EFP items as the following:

- (i) The TDCC should pay the royalty for the block period from 1988-9 to 1990-1 at the same rate they used to pay for the block period from 1985-6 to 1987-8.
- (ii) The royalty for the block period from 1991-2 to 1993-4 should be realised at the same rate for the block period 1988-9 to 1990-1.
- (iii) As regards the block period from 1994-5 to 1996-7 the royalty may be increased by 10 per cent over the previous year of 1991-2 to 1993-4.

4. Market fluctuations / constraints

5. Passing of higher rates to the tribals: The TDCC has been endeavouring to pay higher procurement rates on SAP/EFP as a result of which the cost price exceeds the market price leading to a loss for TDCC.

COMPARISON OF TDCC WITH ITS COUNTERPARTS IN OTHER STATES

The TDCC is not getting the same government assistance like its counterparts in other States. The quantum of financial assistance received by the TDCC from the government *vis-a-vis* its counterparts in other States is given below:

Table 4.8: Financial assistance received by TDCC from the government *vis-a-vis* its counterparts in other States

Nature of assistance	*Annual expenditure of TDCC*	*Assistance received from Govt.*			
		TDCC Orissa	*GCC Andhra Pradesh*	*TDCC Mahara-shtra*	*TDCC West Bengal*
Marginal subsidy	Rs. 1.40 crore	Rs. 20 lakh	Expenditure fully reimbursed	Expenditure fully reimbursed	Expenditure fully reimbursed
Royalty on EFP	Rs. 80 lakh	—	-do-	Loss on trading is compensated	-do-
Purchase tax and sales Tax	Rs. 15 lakh towards purchase tax though collected from the buyers but it affects the sale rate	—	Fully exempted from purchase tax and sales tax		

The TDCC has sought for the following assistance from the government to revamp its financial and business activity.

(i) *Procurement advance*: Annual interest burden of TDCC for undertaking procurement operations is to the tune of Rs. 200.00 lakh. This is a big drain on the scarce resources of the Corporation. It has, therefore, been proposed to the government to sanction a sum of Rs. 8.50 crore as one-time procurement advance in shape of a 100 per cent grant-in-aid.

(ii) Managerial subsidy: The establishment expenditure of TDCC per annum is about Rs. 1.40 crore. Annual assistance of Rs. 20 lakh from the government is not sufficient even for two month's salary of employees of TDCC. The Andhra Pradesh Girijan Corporation and TDCC of West Bengal and Maharashtra get their full managerial expenses reimbursed by their respective governments. Accordingly the TDCC has requested the sanction of managerial subsidy to the

tune of Rs. 1.62 crore to the State Government keeping in view the increase in D.A. payable to the employees.

(iii) Royalty: The TDCC is at present paying royalty to the tune of Rs. 80 lakh to Rs. 90 lakh every year. The TDCC of the neighbouring States are getting it reimbursed from their respective State Governments. In Bihar royalty is exempted for Tamarind.

(iv) Purchase tax and sales tax: On an average the TDCC pays Rs. 15 lakh purchase tax every year. Sales tax which is paid to the government by collecting from the buyers affects the sales trade. The Girijan Co-operative Corporation in Andhra Pradesh is enjoying exemption from payment of purchase tax and sales tax. The TDCC also requests for similar exemptions.

FUTURE TRADE PROGRAMME

The working capital of the Corporation has been completely eroded. The TDCC is not now in a position to carry on procurement of EFP items out of its own fund. In two review meetings taken up by the Principal Secretary-cum-Commissioner, Tribal Welfare Department during October 1992 and January 1993, and by the Hon'ble Minister of Finance on 8.7.94, it has been decided that in view of a resource crunch the TDCC should procure different EFP through tie-up arrangements only and interest burden shall be kept to a minimum. This may affect the procurement operation of the Corporation to a significant extent.

The TDCC collects a number of EFP. Annexure VII gives the collection of EFP by the TDCC during the last 10 years. The procurement chart shows that 17 items of EFP appear to be very important from the point of view of the quantum of collection. These items which exceed 100 qtl. are considered important. These items are Tamarind, Harida, Mahua flowers, Sal seeds, Hill brooms, Nux vomica, Kusum seeds, Marking nut, Mango kernel, Bahada, Amla, Karanja seeds, Kusumi lac, Mahua seeds, Char gum and Sunari bark. Besides these items, Siali leaves and Sal leaves are collected in millions and Puri district is found to be the area from which these leaves are collected in abundance.

The statistics of procurement were maintained districtwise previously. Now these are maintained branchwise. There are 15 branches of TDCC which are distributed all over Orissa.

The analysis of districtwise figures of procurement shows that different districts are important for different EFP. For example, Koraput is famous for Tamarind, Harida, Mahua flowers and hill brooms. Similarly Ganjam is important for Tamarind, hill brooms, Mango kernel and Mahua seeds, Kalahandi district is noted for Harida, Sal seeds and Amla whereas Sundergarh for Kusum seeds, Bahada, Kusumi lac and Char gum, Mayurbhanj for Harida and Nux vomica, and Puri for Nux vomica, Sunari bark, Siali and Sal leaves.

The total value of the EFP collected during the year 1990-1 comes to Rs. 3,76,52,622.00. The value of Tamarind alone is recorded at Rs. 2,47,59,610.80. Next comes Harida of the value of Rs. 48,04,701.00 and Mahua flowers of Rs. 43,55,983.55. Many other items such as Karanja seeds, hill brooms, Siali and Sal leaves, Kusumi lac, Kusum seeds, Nux vomica and Sal seeds also record value in lakhs. Some items like Mahua seeds, Kusum seeds, Neem seeds, Karanja seeds have been transferred from the preview of TDCC and given to a joint sector company. The TDCC was not procuring many items such as broom grass, Tamarind seeds, Bahada gums, thorn brooms, gums, mushrooms, K. creeper, Verisunga leaves, Soap nuts and Sabai grass previously. But these items of EFP are now being procured by the TDCC for the last three years. From time to time new items are collected as and when found available and important, and some items are dropped on account of scarcity and poor marketing outlet.

The TDCC has identified specific items of EFP to be collected in different Forest Divisions of the State in the year 1995-6. The divisionwise list of such items of EFP is given in Annexure X.

CASE ON PRIVATISATION OF TAMARIND COLLECTION

The TDCC used to collect Tamarind from old Koraput district and other areas on the strength of the lease granted to it by the Government of Orissa. Recently the State Government decided to give exclusive rights for the collection of Tamarind in Rayagada and other two branches of Koraput district to a private entrepreneur, Doki Satyanaraya till 2001. This has created unrest in the minds of the workers of TDCC and fear among the tribal people apprehending exploitation by the private lease holder. Tamarind is one of the most important items of EFP which fetches the largest turnover in the State and Koraput is famous for this produce. It is apprehended that there may be serious setbacks in the business of the TDCC. The government's stand for the privatisation of Tamarind collection is that the TDCC is meeting losses in its business year after year and therefore is not in a position to pay royalty to the government regularly.

Utkal Forest Products Limited (UFPL)

M/s. Utkal Forest Products Limited (UFPL), a joint sector company of IPICOL and Mr. J.P. Lath and his associates was promoted in 1989 and the company received lease from the State Government for 10 years commencing from 1989 for systematic collection and processing of essential forest produce in the State of Orissa. (See the Government's letters granting lease to UFP Limited in Annexures XI G & H).

The company has received 27 forest divisions on lease for the collection of 29 items of Essential Forest Produce. In order to have effective collection, the company has appointed a manager (Forest) under whom 3 regional managers have been appointed. Under the regional managers, there are 24 Division-in-Charges who are looking after the entire 27 forest divisions. 125 forest supervisors are working under the supervision of the regional managers.

The site supervisors are receiving the seeds collected by the tribals through collecting agents who have been selected by the company and they are working on commission basis. The tribals sell the produce for the minimum price fixed by the Government of Orissa.

The company is collecting the following seeds i.e. *Ocimum gratissimum* Linn. (bantulsi), *Cassia siamea* Lam. (chakunda), goba, bankulthi, *Acacia concinna* (Willd.) DC. (sikakai), tal makhana, *Martynia diandra* Glox. (baghnakhi), *Holarrhena antidysenterica* Wall. ex A. DC (indrajaba), *Curcuma aromatica* Salisb. (banhaldi), *Mallotus philippinensis* (Lam.) Muell. -Arg. (kamalagundi), marda, *Butea monosperma* (Lam.) Taub. (palash), *Vetiveria zizanioides* (L.) Nash (benacher), *Acacia nilotica* (L.) Delile (babul), *Mangifera indica* Linn. (mango kernel), *Caesalpinia digyna* Rottl. (gillo seed), *Hemidesmus indicus* (Linn.) R.Br. (anata seed), *Mucuna monosperma* DC. ex, Wight (baidanka), *Pongamia glabra* Vent (karanja), *Bassia latifolia* Roxb. (mahua), *Azadirachta indica* A. Juss. (neem), *Schleichera oleosa* (Lour.) Oken (kusum), *Calophyllum inophyllum* Linn. (polang), *Mesua ferrea* Linn. (nageswar) and getting the same processed in its existing solvent extraction/refining plant of Orissa Oil Industries which has its units at Rairangpur and Sambalpur.

Business operations

Business operations of the company for the last 5 years were as under :-

Year	*Qty. collected (in MT)*	*Sales turnover (Rs. in lakhs)*	*Profit/loss (Rs. in lakhs)*
1989-90	3157.47	0.47	(-) 10.73
1990-91	2553.43	23.82	(-) 14.72
1991-92	11534.37	412.30	(-) 20.65
1992-93	5851.37	229.17	(-) 45.62
1993-94	9128.53	587.81	2.23

Initially the company had some problems in collecting the estimated quantity of the forest produce due to financial constraints, etc. However, with the sanction of a higher working capital loan, the collection has improved over the years. The company has also made a modest profit of Rs. 2.23 lakhs after four years of continuous losses.

During March 1994, the company has entered into a MOU with Tribal Co-operative Marketing Development Federation of India Limited (TRIFED) to sell the essential forest produce at the agreed price between the company and TRIFED. TRIFED is getting the material processed in the plant of Orissa Oil Industries Ltd. located at Rairangpur and Sambalpur.

Rs. 2.3 lakh project was sanctioned by IPICOL and OSFC with the following parameters.

Project		*Cost (Rs. in lakh)*
(i) Plant & machinery	:	15.80
(ii) Auxiliary equipments	:	78.00
(iii) Miscellaneous fixed assets	:	8.00
(iv) Contingencies	:	5.09
(v) Preliminary & preoperative expenses	:	2.00
(vi) Margin for working capital	:	94.11
Total	:	**203.00**

Means of Finance (Rs.in lakh)

Share capital

(a) Private promoter	:	51.00
(b) IPICOL	:	48.00
(c) Unsecured loan (Promoters)	:	24.00

Term loan :

(a) IPICOL	:	40.00
(b) OSFC	:	40.00

A. Name of organisation : M/s. Utkal Forest Products Limited, Ainthapali, Sambalpur.

B. Incorporation : The Company incorporated in the year 1989 and started production since 1990.

C. Organisational set-up : The Company is having its head office at Ainthapali, Sambalpur and various regional office/ divisional offices throughout the state. The company follows a flat organisational structure with centralised administrative system. The organisational set-up is as follows:

I. *Head Office* : It is the Chief Administrative Office of the Company which controls the entire activities of the organisation such as, administration, finance and marketing.

II. *Regional & Divisional Office* : The Regional & Divisional Office is responsible for collection of various EFP for the divisions and co-ordinates with divisional forest offices of the Govt. of Orissa. All the divisional offices have been lowered down to various ranges and rests finally in collection centres.

III. *Marketing Office*: The Company has also set-up a marketing office at Calcutta which is looked after by the Managing Director.

The organisational chart of the company is depicted as follows :

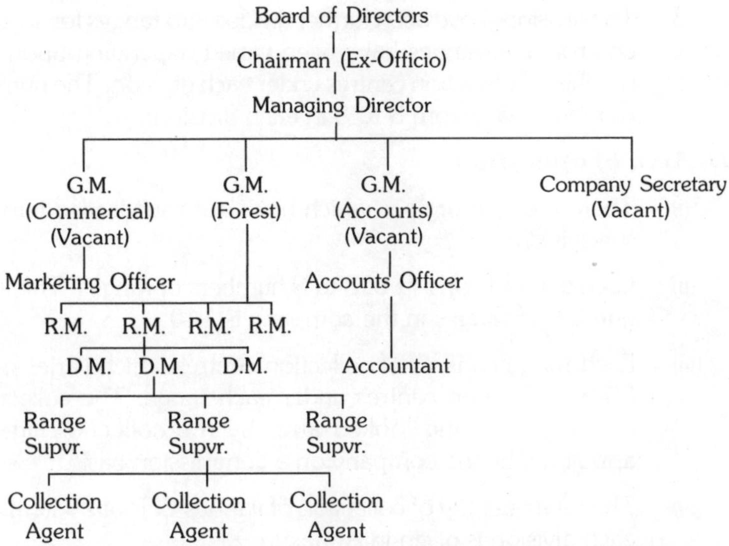

1. The Managing Director looks after the entire Administration, Production, Finance and Marketing. He is assisted by G.M. Forest in respect of collection of EFP. In absence of G.M. (Account) and G.M. (Commercial), the Managing Director is being assisted directly by the lower level officers in accounts and marketing.
2. The entire forest area of Orissa has been regraded into four regions and 27 divisions which are as follows:

Regions	*Divisions*
1. Region 1	Baripada, Karanjia, Keonjhar.
2. Region 2	Sambalpur, Bamara, Deogarh, Sundergarh, Bonai, Bolangir, Angul, Rairakhol, Athamalik, Dhenkanal, Athagarh.

3.	Region 3	Nawarangpur, Jeypore, Rayagada, Baliguda, Kalahandi, Khariar.
4.	Region 4	Parlakhemundi, Phulbani, Boudh, Nayagarh, Ghumsar North & Ghumsar South.

3. The divisions have been further divided into ranges for smooth control. The ranges have been varied depending upon the number of collection centres under each division. The number of ranges vary from 3 to 8 in each division.

D. Area of operation :

(i) There are four regions which have been subdivided into 27 subdivisions.

(ii) Each division consists of 3 to 8 numbers of ranges. The total number of ranges in the company is 140.

(iii) Each range consists of collection centres which varies from 50-60 collection centres under each range. The collection centres are being looked after by the collection agents appointed by the company on a commission basis.

(iv) The total quantity of collection of various EFP of 29 items, in each division is given in Annexure XI.

(v) The collection season which is by known as forest season is September-October of each year. The peak is from April to September and November to January.

E. (I) Processing units

The Company does not have its own processing units for processing of EFP collected by UFPL The Company processes the item, so collected (having oil bearing contents) through its sister companies namely :

M/s. Orissa Oil Industries Limited having 2 processing units (a) At Sasan and (b) at Rairangpur.

M/s Utkal Oil Limited having its processing plant at Ambaguda, Koraput.

(II) Items processed

The following oil seeds are being processed in the above-mentioned processings units.

(i) Mahua, (ii) Mango kernel, (iii) Neem, (iv) Karanja, (v) Gillo seed, (vi) Tulsi, (vii) Chakunda, (viii) Banakulthi, (ix) and other oil-bearing seeds.

The seeds which do not have oil-bearing contents are sold directly by the company.

F. The details of EFP collected and processed, divisionwise collection of EFP, and quantity sold during last three years are given in Annexure-XI (A), (B), (C), (D), (E) & (F) as detailed below:

XI-A Statement showing the quantity of EFP procured and the quantity processed by the Utkal Forest Products Ltd. during 1992-95.

XI-B Statement showing processiong cost, total collection, and sale value of EFP collected by UFPL during 1992-93 to 1994-95.

XI-C Divisionwise collection of EFP items processed by UFPL from 1.10.92 to 30.9.93 (in quintals).

XI-D Divisionwise collection of EFP items by UFPL for the period ending 30.9.94 (in quintals).

XI-E Divisionwise collection of EFP items by UFPL from 1.10.94 to 30.9.95 (in quintals).

XI-F Quantity of EFP sold by Utkal Forest Products Ltd. from 1992-3 to 1994-5.

Large size Multi-Purpose Co-operation Society (LAMPS)

The tribal communities were facing difficulties in selling their SAP and EFP and also getting their daily necessities or consumer goods. The local traders who were fulfilling these two needs were exploiting by charging a high price for the consumer goods and giving a very low price for their SAP and EFP. In order to solve this problem the Bawa Committee recommended that LAMPS should operate in the tribal areas to buy their produce and supply consumer goods to them. In pursuance of this recommendation, the State Government has established 223 LAMPS each covering 10,000 to 20,000 population in the tribal areas.

Several LAMPS have been attached to the TDCC and also to the OFDC for procurement of EFP on their behalf and many of the LAMPS are also carrying on their business in EFP independently.

As the situation stands today it shows that the LAMPS have failed in achieving the objectives for which they were set-up. The following structural limitations need to be tackled for the LAMPS to operate effectively:

(i) The large areas covered under the LAMPS make them little effective to meet the demands of the tribals. The population is so scattered that quite often it takes a whole day for some members to come to the LAMPS on foot.

(ii) Heavy transportation expenditure is involved in carrying goods from purchasing centres to branch offices/head offices located in almost inaccessible and remote places in the hilly areas.

(iii) The financial condition of the LAMPS is very weak. They are unable to purchase the required quantities of fertilizers in time to meet the demands of the farmers.

(iv) Most of the LAMPS are running at a loss to such an extent that they are not in a position to pay even the salaries of the staff members.

(v) The transport charges paid by the Orissa State Co-operative Marketing Federation Ltd. is not sufficient to meet the expenditure incurred by LAMPS towards transportation of fertilizers.

(vi) Commission paid to LAMPS is not very encouraging. After deducting transport costs and handling charges the margin left is negligible.

(vii) LAMPS incur loss as shortages of fertilizers occur in retail sales. Sometimes large shortages of fertilizers are incurred by LAMPS due to improper and prolonged storage. Most of the LAMPS have no improved godowns, which also results in damage to the stocks during storage.

(viii) Quite often the official hours of operation of the LAMPS are not convenient to the tribals. As a result, both clients and business suffer.

(ix) The LAMPS are supposed to act as agents of the Tribal Development Co-operative Corporation of Orissa Ltd. (TDCC) in procuring EFP. But, the TDCC does not appoint all the LAMPS as its agents and all those appointed are not given agency to procure all types of EFP. The tribals are not approaching the LAMPS for sale of their produce for which they are primarily meant and organised. This is mainly because of lack of co-operation and co-ordination among LAMPS, TDCC and other co-operative agencies.

(x) Co-ordination and healthy organic linkage between different tiers of co-operative marketing structures (both horizontal and vertical), viz., LAMPS, TDCC, RMCS, OSCMF, TRIFED, NAFED are lacking. Sometimes unhealthy competition among different co-operative organisations is discernible.

(xi) Management and supervision of advances/loans, and their recovery need further streamlining.

(xii) Arrangement for linking credit with marketing of SAP and EFP needs to be strengthened.

(xiii) The share capital and managerial assistance is not adequate and hence most of the LAMPS borrow funds from the financing agencies to undertake consumer business. The financing agencies charge very high rates of interest, to the extent of 13 per cent and above. Due to little margins and high managerial expenses, the LAMPS are not able to repay the bank loans and their viability is affected.

(xiv) The staff members of LAMPS do not belong to a single cadre. They are brought on deputation basis from Central Co-operative Banks, Co-operation Departments, TDCC, etc. Even for the staff of the same ranks, different scales of pay are paid. The salary variations among the staff affect the work efficiency to a large extent.

The Agency Marketing Co-operative Society Ltd. Tikabali

The Agency Marketing Co-operative Society of Tikabali was conceived just after Independence. Within a short period of its origin, it came into prominence. The quality and sincerity of work of the society placed it as a leading example in the country. It originated with only

802 members and now it boasts of a large family of members of 54,000. It has not only increased the number of members, the wage level of the workers has also been steadily raised. The quantum has multiplied many more times and the society has extended its hands in a large area of activities.

The society is a lessee of the Department of Forest, Government of Orissa. It collects 18 items of EFP in 3 Forest Divisions namely Baliguda, Boudh and Phulbani of Kandhamal District (old district) through its tribal and Harijan members. The items are Siali plates, Tamarind, marking nut, arrowroot, brooms, etc. It also procures oil seeds, pulses from the cultivators successfully eliminating middlemen. To add to this, the society provides daily necessities, controls textiles and commodities, and baby foods to thousands of tribal and Harijan members through its 74 collection centres are spread over every nook and corner of the district at a reasonable price.

The society is maintaining a college at Tikabali where 400 students are getting opportunity for higher education. In addition to this the society is giving financial assistance to other private colleges and schools for the construction of classrooms, purchase of library books, etc., for the education of tribals and Harijans of the district.

The head office of the AMCS college is located at Tikabali whereas its zonal and branch offices are located at Baliguda and Berhampur respectively. At present it has 53097 members consisting of 45119 ST, 7924 SC and 53 other caste members and one State Government representative.

The Society has working capital of Rs. 330.69 lakh and the share capital amounts to Rs. 22.78 lakh. Items of EFP leased out by the State Government in favour of AMCS for 1995-6 and 1996-7 are given below :

1. Siali leaves and plates
2. Hill brooms
3. Tamarind
4. Matreeds
5. Marking nuts
6. Arrowroots
7. Honey and Wax
8. Lac
10. Genduli gum
11. Myrobolan
12. Babul gum
13. Sabai grass
14. Thatching grass
15. Cane
16. Nux vomica
17. Simuli cotton

9. Cleaning nut   18. Siali fibre

The AMCS deals with the procurement and sale of EFP and consumer goods and textiles. The following figures show the amount of transactions during last three years.

Year	*Purchase (Rs. in crore)*	*Sale (Rs. in crore)*
1992-3	1.90	2.91
1993-4	2.28	3.62
1994-5	11.88	3.10

Tribal Co-operative Marketing Development Federation of India Ltd. (TRIFED)

One of the greatest problems which the tribal communities face is their exploitation through extraction of usurious rates of interest, fraud in purchase and sale, payment of wages much lower than the statutory prescribed minimum debt bondage, liquor vending and in various other forms. Various measures which are being taken and various legal enactments and regulations which have been framed under the Fifth Schedule are meant for protecting them from exploitation. Credit and marketing solve both protective as well as developmental measures.

Out of this idea has emerged the TRIFED. The need for a Central Apex Level Federation like TRIFED was long felt and was also recommended in successive meetings and conferences and also by a working group on development of scheduled tribes during the Sixth and Seventh Plan Period. The Government of India accepted the recommendation of the working group and set-up TRIFED as an apex body at the National level of the State level Tribal Development Corporation, Forest Development Corporation and other such State bodies in August 1987. Broadly the objectives of TRIFED are as follows:

(i) Full utilisation of natural resources of tribal areas by improving their marketability.

(ii) Higher earning and employment opportunities for the tribal people.

(iii) Creation of awareness regarding the market forces among the tribal people to enable them to optimise their income.

(iv) Marketing support to the State tribal co-operative corporations and forest corporations engaged in such activities.

In order to make the TRIFED effective in their operations, two conditions are to be fulfilled. One is to bring more items of EFP in the nationalised sector and the other one is to abolish royalty being levied on EFP. The government have so far nationalised two items, namely, Sal seed and Kendu leaf. Sal seed has been nationalised since 1983, and OFDC & TDCC Ltd. are entrusted with the collection of Sal seed in different divisions which change from year to year. Kendu leaf has been nationalised since 1973. The Forest Department has been entrusted with the collection of Kendu leaf, and OFDC Ltd. with the marketing of the same. The royalty is paid by the agencies entrusted with the collection for both the items.

The TRIFED was to conduct training of the tribal people in scientific collection of various items of forest produce and to take steps for the strengthening of the LAMPS so that these become viable collection agencies. The function of arranging training has not been exercised by the TRIFED. The tribal communities go on collecting various EFP according to their traditional methods of collection. Some of these procedures are extremely injurious to the plants. For example, the practice of tapping Gum Karaya is so harmful that the number of such plants has now dwindled to a great extent and the quantity of gum now collected has become so small that the produce has lost its marketing potentiality. TRIFED purchases various items of EFP largely from the TDCC and UFP Ltd. and gets the oil-borne seeds processed by Orissa Oil Industries and Utkal Oils Ltd. Similarly Mangula Traders and S.N. Rice Mill also do processing of certain EFP received from TRIFED. Annexure XII gives figures of procurement of SAP and EFP by TRIFED from 1988-9 to 1994-5.

Orissa Rural Development and Marketing Society (ORMAS)

ORMAS was created by the Government of Orissa in 1989 as an autonomous State Agency under the Department of Panchayati Raj, Government of Orissa. It is meant to create an institutional outlet at various levels (State, District and Block) for the marketing of rural produce at a remunerative price. It is also meant for playing a leading role in the mobilisation and organisation of the unorganised rural poor, the formation of indigenous groups, imparting new skills and upgrading existing skills and to help in the process of creation of a trained reservoir of manpower in rural areas. Basically ORMAS aims at bridging the critical gap in the chain from production to marketing especially for the beneficiaries of Integrated Rural Development Programme (IRDP).

In Orissa there are about 3.28 lakh beneficiaries who have been assisted in the non-farm sector under IRDP till the end of the Seventh Plan. Out of them only about 40,000 have been covered so far by existing marketing agencies in the state. The rest of the beneficiaries have been left to seek their own fate without any marketing support from any agencies. The ORMAS is born to take initiative in using modern scientific marketing techniques and approaches for helping the poor rural people in the marketing of their product. Value addition through marketing for increasing the income of poor rural producers is the basic aim of ORMAS's operation philosophy.

During the last five years, ORMAS has launched a number of initiatives to fulfil its mandate. It has set-up District Supply and Marketing Societies (DSMS) in all the divided districts of Orissa. Orissa is the first state in India to accomplish this professionalisation of DSMS in all districts. It has organised a number of fairs and exhibitions in the State and also has participated in similar fairs and exhibitions organised by CAPART to give a boost to rural producers and in particular the rural artisans and DWCRA groups. One of the important contributions of ORMAS in tribal development is to assist members of the DWCRA group in marketing EFP which they collect from the nearby forest including sincere endeavour for value addition through preliminary processing of the produce at their doorsteps. In the meantime, the government has taken a decision that the DWCRA groups be allowed to collect EFP from their respective localities and do the necessary preliminary processing for value addition. ORMAS will provide necessary support for marketing of the finished products of this group at remunerative prices. Necessary royalty and commission octroi shall be paid to the TDCC by the ORMAS on behalf of the DWCRA groups.

This practice has created trouble in certain areas. The TDCC considers the collection of EFP and the processing of these items by the DWCRA groups as an act of unlawful encroachment into the area over which they have monopoly as lease holder. Out of this situation have emerged some cases of conflict between the DWCRA groups on the one side, and TDCC and Forest Department on the other. A case of such conflict is given below.

HILL-BROOMS CASE STUDY

The Mandibisi Mahila Mandal located at Mandibisi village, Kashipur area, procured hill-brooms for processing and preparing broomsticks. The TDCC is the licence holder of this area as regards collection and sale of hill-brooms. Therefore, the collection of hill-brooms by the Mahila Mandal is considered unauthorised procurement. However, in the collector's conference, it was decided that the Mahila Mandal Samiti should be authorised with the collection of EFP. In line with this decision, the TDCC decided to appoint the Mahila Mandal Samiti as a sub-agent of TDCC. In the meeting presided over by the Chief Secretary of Orissa, it was decided that ORMAS will function as the nodal agency to sponsor the application from women groups to TDCC for appointment as sub-agents in their respective areas. Accordingly, ORMAS sponsored the application of the Mandibisi Mahila Mandal who refused to become sub-agent under TDCC on approach to the Committee by the Branch Manager of TDCC, Kashipur and Divisional Manager of TDCC, Sunabeda.

It was reported that the Mandibisi Mahila Mandal procured 173.90 qtl. of loose hill-brooms at the rate of Rs. 400 per qtl. from the tribal collectors whereas the approved rate fixed by TDCC was Rs. 500 per qtl. As the procurement was unauthorised, the Branch Manager of TDCC complained to the Forest Division, Rayagada, against the Mahila Mandal and in the year 1995 the Forest Mobile Squad consisting of a Forest Ranger, Police Staff, the Tahasildar-cum-Executive Magistrate, the Branch Manager TDCC, Kashipur and some of the local staff of the Corporation detected and seized the hill-brooms. The mobile squad first persuaded the Mahila Mandal to hand over the brooms to the TDCC at the rate of Rs. 500 per qtl. as the Corporation is the lease holder. They refused to hand over the same. The Forest Ranger seized about 69.26 qtl. of loose hill-brooms and handed them over to the Branch Manager, TDCC, Kashipur for safe custody. In the same year, one forest case and one police case were filed against the Mahila Mandal. The Mandibisi Mahila Mandal filed a counter F.I.R. against the Forest Ranger, Kashipur and Branch Manager, TDCC for committing alleged atrocity on them. Moreover, the Mandibisi Mahila Mandal filed a writ petition before the Hon'ble

High Court, Cuttack, Orissa that they have collected the stock of hill-brooms in good faith in anticipation of getting a licence and Corporation should purchase the stock at the rate of Rs. 700 per qtl. The writ petition filed by the Mahila Mandal was finally heard and disposed by the Hon'ble High Court, Cuttack, Orissa. As per decision, the TDCC has requested the Welfare Department to ask the Forest Department to dispose of the stock and deposit the cost to the TDCC Divisional Manager, Sunabeda.

Manner of procurement

The lease holders engage agents or Thikadars to remain in charge of collection of EFP of specified areas. The agents in turn appoint local tribals to watch the EFP so that no one collects the EFP by stealth. The agents or those who have been appointed to remain in charge of the collection as sub-agents, go round periodically in their respective base areas to ensure vigilance of the watchmen. Wherever vigilance is lacking some business men influence the tribal people and arrange the collection stealthily.

There are collection centres set-up at several places. Some are permanent and some are temporarily opened. The collectors carry the produce to such centres and sometimes the agents procure the produce from the collectors on weekly-market days. The products collected at the collection centres are transported from where the products are lifted by the purchasers whether they represent co-operative organisations, corporation or private agencies.

Storage facilities are lacking at the field level. A place is cleaned and the collected-stuff is dumped at the place under a tent. After the collection is over, the items are packed in gunny bags which are transported by bullock carts to the roadside, and from there transported to the godowns by trucks.

PRICING PROCEDURE AND QUANTITY OF EFP COLLECTED

Price fixation committee

In every district, there is a Price Fixation Committee headed by the Collector for fixation of procurement price of various items of EFP available in the district. The Committee consists of the following members:

1. Collector	Chairman
2. Deputy Registrar, Co-operative Society	Member
3. D.F.O., Forest Department	Member
4. Branch Manager, TDCC/OFDC	Member
5. Assistant Registrar, Co-operative Society	Member-Convenor

Besides these members, the Managing Director/President of local LAMPS and Co-operatives the Managing Director, Utkal Forest Products Ltd. and some other officials are invited to attend the meeting. The purpose of the formation of the Committee is to safeguard the economic interest of the tribal collectors by ensuring to them a reasonable price and to save them from exploitation by unscrupulous traders. A minimum price is fixed by the Committee and the TDCC, LAMPS and other agencies engaged in procurement of EFP make payments according to the price fixed by the Committee and approved by the concerned Revenue Divisional Commissioner. The price list after approval is notified in the collection centres. Announcements are also made in the weekly market by the TDCC/OFDC staff.

FIXATION OF SALE PRICE BY TDCC

The EFP items procured directly from the collectors and from weekly markets are kept in godowns located in different branches after intimation to the Head Office of the TDCC. As regards the sale price, the TDCC is the final authority to fix up the price except of the nationalised item like Sal seed, the rate of which is fixed by the Forest and Environment Department. While fixing up the sale price of different items, the following costs are taken into account.

1. Average procurement price
2. Royalty
3. Purchase tax/Sales tax
4. Octroi/RMC
5. Cost of gunny bags
6. Godown rent
7. Packing, forwarding, transportation and handling charges.
8. Driage/shortage

9. Establishment charges
10. Bank interest
11. Margin/profit

The costing of Tamarind and Harida for example is given in Table 4.9. The EFP items are usually sold to TRIFED and other private agencies which lift the required items from the godowns at the branch level.

Table 4.9: Pricing of Tamarind and Harida by TDCC (in Rs.)

		Tamarind (*per qtl.*)	*Harida* (*per qtl.*)
1.	Procurement price	300.00	150.00
2.	Royalty	10.00	20.00
3.	Purchase tax/sales tax	12.00	6.00
4.	Octroi/RMC	6.00	3.00
5.	Gunny bags	10.00	16.00
6.	Packing, forwarding, transportation and handling	25.00	25.00
7.	Godown rent	—	—
8.	Driage/shortage	15.00	Nil
9.	Establishment charges	—	—
10.	Bank interest	Not applicable	Not applicable
11.	Margin 15%	45.00	22.00
		423.00	242.00

Quantity of EFP

EFP available in the forests of Orissa are of many kinds, of which only about 101 items have been identified and given lease to different agencies for procurement. The categorywise classification of EFP, the area where they are available in abundance, season of collection, average quantity of items procured yearly, etc., have been given in Annexure V. But it does not give the picture in totality. The volume of procurement by such agencies as TDCC, OFDC, and UFP finds place in this Annexure. There are, however, many private agencies and co-operative societies and LAMPS who also collect EFP. Therefore, it is necessary to give a total picture of the collection of EFP. Annexure XIII gives information about the volume of 65 items of forest produces collected from the year 1985-6 to 1991-2. The Annexure shows that among the EFPs, Kendu leaves top the list in respect of quantity of collection followed by Mahua flower, Sal seeds and Tamarind. So far as the value is concerned, Kendu leaves come first giving the maximum return to the government. Next come Mahua flowers followed by Sal seeds, Tamarind and Genduli gum in order of gross value in rupees.

The production of Kusumi lac which reached a peak level of 1381 MT in 1967-8 declined to 209 quintals in 1989-90. Another important EFP, namely gum, has met the same fate. There was a time when gums were extracted from various tree species, but there has been not only a reduction in quantity of collection of gums but some of the species are in the process of vanishing from the forests.

Even though plants and trees containing tannin are abound in the forests tanneries are just a few in Orissa. Most of the barks and other parts of the plants and trees with tannin content are sold to tanneries located outside the State.

The state is a huge storehouse of tree-borne oilseeds starting from Sal seeds or Mahua seeds which contain components of Cocoa butter in fats to Chakunda and other seeds containing oil of medicinal value. These oils are of great value and of multiple utilities. These oilseeds are procured in multiple metric tonnes and have prospects for the eradication of poverty among the tribal people who are the sole collectors of these seeds. Very little has been done in the matter of processing and utilisation of oil and fat for various purposes. The Orissa forests are rich in medicinal herbs and aromatic grasses. Very little work has been done to extract alkaloids and other ingredients of medicinal value from the medicinal herbs. Similarly there is a vast scope for cultivation of aromatic grasses and spices. Even though the technical know-how is available with RRL, Bhubaneswar, very little progress has been made in the cultivation of such grasses and extraction of essential oils which are in great demand for export to foreign countries.

Potentialities of production of all types of EFP can be determined by various survey and inventory methods. But no such survey has been done. The figures collected from official records present a small fraction of the potential availability. Moreover, the collections by clandestine practices and pilferage are unknown and the quantum of EFP which rot and get decomposed lacking collection in time is beyond any estimation. Taking all these facts and figures into consideration, the productivity and availability of EFP may be many times of what the official sources speak of.

Royalty

The National Commission on Agriculture reported that the visible contribution of EFP to the total revenue of the Forest Department in Orissa was 28.9 per cent in 1969-70. With the discovery of new uses of many EFP, the revenue has increased considerably. The details of revenue in rupees for the years 1991-2 and 1992-3 are reproduced below from the brochure, Orissa Forests, 1993, published by the Office of the PCCF, Orissa.

Details of revenue 1991-2 & 1992-3:

Items	*Revenue (in % of total forest revenue)*	
	1991-2	*1992-3*
Kendu leaves	Rs. 51,57,32,000 (60.91%)	Rs. 78,25,74,602 (75.31%)
Other MFP(EFP)	Rs. 5,44,11,556 (6.42%)	Rs. 1,93,65,280 (1.86%)

The amounts of royalty paid for Kendu leaves by the OFDC and against all other EFP by other agencies for the years from 1980-1 to 1992-3 are given in Table 4.10, and Table 4.11 gives information regarding royalty paid by different agencies against Sal seeds, Kendu leaves and other EFP for the last two years — 1993-4 and 1994-5. The royalty paid to the government by the OFDC for the year 1993-4 in respect of Kendu leaves was Rs. 7434.93 lakh and in 1994-5 for a part of it was Rs. 4869.37 lakh.

Table 4.10: MFP(EFP) Royalty from 1980-1 to 1992-3 (Rs. in '000)

Year	*Kendu leaves*	*Other MFP(EFP)*	*Total*
1980-1	88804	13989	102793
1981-2	90000	17591	108251
1982-3	140061	40316	157652
1983-4	185000	40316	225316
1984-5	17000	35892	205892
1985-6	130000	46080	176080
1986-7	156949	45904	202853
1987-8	274828	36737	311565

1988-9	236382	23991	260373
1989-90	824444	30666	855110
1990-1	768500	20510	789010
1991-2	515732	54412	570144
1992-3	782575	19365	801940

Table 4.11: MFP(EFP) royalty paid by different agencies (excluding private agencies) during the year 1993-4 and 1994-5

Sl. No.	*Name of Agency*	*Royalty amount (Rs.in '000)*		*Remarks*
		1993-4	*1994-5*	
1.	Tribal Development Co-operative Corporation of Orissa Ltd.	1181	2748	Royalty for Sal seeds and other MFP(EFP) including arrear Royalty.
2.	Orissa Forest Development Corporation Ltd.	753493	486937	Royalty of Sal seeds and Kendu leaves.
3.	Utkal Forest Product Ltd.	2813	2929	
4.	Agency Marketing Co-operative Society Ltd., Tikabali	1585	1714	
5.	Forest Produce Marketing Co-operative Society Ltd., Mujagada	127	139	
6.	Gunpur LAMP	150	165	
7.	Kunduli LAMP	366	403	

Two agencies namely the OFDC and TDCC deal with the collection and marketing of Sal seeds. Royalty paid to the government in respect of Sal seeds by the OFDC is given in the Annexure III(A) for the last twelve years from 1983 to 1994. The figures show an erratic fluctuation starting with Rs. 11.32 lakh in the year 1990 to Rs. 420.50 lakh in the year 1991. This erratic trend may be due to the fluctuating quantity of production which is a characteristic feature of this item of EFP. Otherwise compared with other items, the Sal seed is an important revenue-generating EFP.

The Utkal Forest Products Ltd. which has been given lease for the collection of 29 items of EFP, all of seed types, pays a sizeable amount of royalty to the Forest Department of the Government of Orissa. In 1992-3, the amount was Rs. 2518920.00 which increased to Rs. 2813185.00 in 1993-4. The royalty has further increased to Rs. 2928805.10 in the year 1994-5.

The Agency Marketing Co-operatives Society Ltd., Tikabali has been given lease to collect 18 items of EFP in three forest divisions — Baliguda, Boudh and Phulbani. It pays royalty to the government for the items which are included in its business activities. The royalty paid in the past three years is mentioned below:

1993-4	Rs. 15,85,284.00
1994-5	Rs. 17,14,296.00
1995-6	Rs. 19,01,325.00

The TDCC, which deals with many items, does not maintain up to date figures of royalty paid to the government itemwise. However, the records show that this organisation has paid royalty.

Year	*Royalty paid (in Rs.)*	*Remarks*
1985-6	12,12,706.00	
1986-7	12,12,706.00	
1987-8	12,12,706.00	
	1,14,787.00	(Arrear Royalty)
1988-9	12,12,706.00	
1989-90	11,41,839.00	
1990-1	12,12,706.00	
	42,350.00	(Arrear Royalty)
1991-2	12,12,706.00	
1992-3	12,12,706.00	
1993-4	11,81,226.00	
1994-5	19,71,864.00	
	7,76,106.00	(Arrear Royalty)

Tribal communities and exploiting elements

In order to improve the economy of tribal communities and protect them from exploitation by the local merchants, moneylenders and middlemen, the BAWA Committee recommended the organisation of Large size Agricultural Multipurpose Co-operative Societies (LAMPS) at the primary level. The objective was to provide production as well as consumption credit, to undertake the marketing of agricultural and EFP, to supply agricultural inputs and distribute consumer goods so that the tribals get all their requirements from one agency.

Collection of EFP in different seasons is one of the important occupations of the tribals. About 30 to 50 per cent of the income is obtained through the sale of EFP, the rest being derived from agriculture. The tribals are being exploited by private traders in their sale transactions of EFP. The lease of EFP is given to OFDC, TDCC, LAMPS and various other agencies including private agencies for the procurement of EFP. Thereby the private traders are eliminated from the field. In principle the TDCC works as the main lessee and LAMPS act as their agents of procurement. In some cases where the lease is granted to a LAMPS it appoints another LAMPS which is situated nearest to the forest area as its own agent of procurement for the area concerned. As per the present arrangement, the TDCC takes the lease of the EFP from the Forest Department and LAMPS collect the forest produce on behalf of TDCC in the capacity of their agents and supply to the TDCC all the items collected by them as per terms and conditions mutually agreed.

In order to prevent the exploitation of tribals, the government has empowered collectors to fix minimum procurement price of different items of EFP. The members of the Price Fixation Committee constituted by the local officials belonging to the Forest Department, co-operative societies including LAMPS, and Government of Orissa undertakings such as OFDC and TDCC give advice to the Collector who is Chairman of the Committee regarding the price to be fixed for each item of EFP. During the procurement seasons, the LAMPS are expected to maintain close contact with TDCC and bring to their notice any fluctuation in prices, availability of produce, etc. The TDCC is expected to review the whole situation and give directions to the LAMPS in these matters. These procedures are, however, rarely followed.

The Branch Manager of LAMPS or their authorised agents purchase EFP from the tribals whenever they bring it either to the office or to the depot or to the weekly market. The salesman is supposed to put up the price list on a board at all these places for information to the tribal sellers. The salesmen are supplied with a sufficient number of balances and weights and measures of all denominations, and use them in all buying and selling transactions.

In many tribal areas the barter economy is still prevalent. This is a medium through which the tribal people are cheated by private merchants. The barter transactions are allowed at the LAMPS level. On this basis, all commodities brought by the tribals for sale, are accepted by the LAMPS and the tribal sellers are remunerated to the extent of their requirements of the value of EFP bartered by them. It is expected that the persons of the LAMPS who are in charge of business transactions, are supposed to explain the correct weights and measures, and inform the rates of different commodities to the tribal people during buying and selling transactions with them.

Certain items of EFP are collected in huge quantities, which require storage under proper conditions. It is a normal duty of the LAMPS to see that the godowns are kept in readiness before the procurement season starts. It is the common understanding that in reality there should be no private traders in the field to purchase EFP items from the tribal people. In such a situation, the question of the LAMPS purchasing from the private traders does not arise. Moreover, Purchase Registers are maintained to keep record of names and addresses of persons from whom items are purchased. Strict checking is supposed to be done by higher authorities. It is a rule that whenever anything is purchased from the tribals, full payment should be made. No deferred payment is to be made.

The Managing Director of the LAMPS is supposed to assess the expected volume of different items of EFP to arrive on the Haat (weekly market) day during the peak procurement season. All arrangements relating to money, coins, packing materials, weights and measures, additional hands are supposed to be made in advance so that all items of EFP in the fullest quantity are procured. All precautions have to be taken so that any breakdown in the middle of the operation should not happen.

From time to time various other measures are being taken to check the exploitation of the tribal communities in the forestry sector relating in particular to the payment of fair prices for the forest products including EFP. Following the recommendations of the National Commission on Agriculture, 1976, the conference of Ministers in charge of Forest and Tribals, 1978 and the National Committee on the Development of Backward Areas (NCDBA), 1981, the State Government has paid considerable attention to the departmental working of forests and nationalisation of trade of forest produce namely Kendu leaves and Sal seeds with a view to checking exploitation of tribals and ensure a fair price for primary tribal collectors. To make the anti-exploitative measures effective, a Central Apex Level Federation namely the Tribal Co-operative Marketing Development Federation of India Ltd. (TRIFED) has been set-up and it is functioning since August, 1987. Now it is to be seen to what extent have these various measures been effective in eliminating the middlemen and strengthening the economic base of the tribal people. With the realisation of commercial importance and with a view to eliminating middlemen in the business, the Kendu leaf trade was nationalised and the working has started since January 1973. In spite of wholesale nationalisation of this trade it has not yet been possible to make the real bidi manufacturers take part in the auction sale held at Sambalpur. The same purchasers and agents still actively continue as middlemen in the trade and they are the ones who attend the auction sales and get the benefit as they used to do before. In fact, the results of nationalisation have rather worked opposite to the objectives of achieving economic development for the tribal communities. As a result of this measure the number of legal buyers has been reduced. The free flow of goods has been hampred, and payment to the gatherers has been delayed. In such a situation, it has been easier for the contractors and middlemen to appear in the scene through a backdoor. Ultimately, this results in the reduction in the collection of EFP and income of the tribal collectors.

A monopoly system is in operation in the State. The Orissa Forest Act confers this right to the State Government by virtue of which the State exercises monopoly over any forest produce earmarked for this purpose. For example, the UFP exercise monopoly purchase rights over 29 items of EFP, and AMCS over 18 items in particular areas. The idea is to protect the tribals against exploitation by private traders and middlemen. Enhancement of revenue and fulfilment of the demands by industrial organisations are also mixed with the objective of protection of the interest of the tribals. What has happened in the long run is that the value addition at the lower level has been completely ignored, protection of the interests of the tribal communities has been downgraded, and the increase in revenue is given the topmost position in the hierarchy of objectives. In this context, it may be mentioned that one does not understand what justifications are there to give the power of monopoly procurement over 29 items of EFP to a private trader. In spite of our best efforts, even IPICOL who has collaborating links with the trader, could not provide procurement prices of different items of EFP over which the concerned trader exercises the power of monopoly procurement.

Two major institutions have been set-up by the State. One is the OFDC and the other is the TDCC. Both of them are in default. The TDCC is confronted with a massive liability of Rs. 1871.11 lakh which is to be paid to the banks and other organisations for the clearance of the loan, and after the loss to the tune of about Rs. 20 crore in the business of Mahua flowers, the OFDC has become almost a pauper. It has been a difficult affair on the part of both the organisations to maintain the huge manpower base of their establishment. In fact, the TDCC runs its business and maintains its establishment on advance payment received from its agents who have been employed to procure particular items of EFP of which the TDCC is the authorised lease-holder. With the privatisation of the Tamarind trade in areas like Rayagada which is the greatest producer of Tamarind, the TDCC, is sure to suffer the greatest setback. These state of affairs clearly show that the State monopoly has encouraged private monopoly to come up with resultant adversities with particular reference to payment of remunerative prices to the tribal collectors.

One or two concrete examples may be given to throw more light on the marketing and trading imperfections resulting from monopoly trading and lack of vigilance on the part of the government.

KENDU LEAVES

Kendu leaf trade is under wholesale nationalisation. No private person is entitled to collect the leaves. There are areas where clandestine collection of Kendu leaves are going on, and private traders and bidi manufacturers make use of these leaves collected illegally by the local pluckers, and earn huge profits with little gain to the pluckers as well as to the government.

HILL-BROOMS

It is the law that the lease holders like TDCC and its traders can process the hill-brooms to make broomsticks and the tribal collectors are not permitted to do the processing of this product. Not only to the tribals it is forbidden to do the bindings of the brooms but they are also refrained from selling the collected items in the open market. Thus the tribal people are deprived of processing the item for value addition and disposing of the produce in the market of their choice to get a better price. The tragic instance of a DWCRA woman group of Rayagada region is a case in point. For details see the Telegraph, 11 June 1995. The women members of DWCRA were prosecuted and the stock of hill-brooms which were collected by them was seized and it finally resulted in a huge financial loss to the organisation.

Likewise there are many such instances which adversely affect not only the economic base of the tribal people but also degenerate many EFP yielding species. There is no item of EFP which is collected by the tribal people and which cannot be processed at their level for value addition. It is a well-known fact that the National Forest Policy, 1952 completely ignored the symbiotic relationship between the tribals and forests. This insensitivity of the National Forest Policy 1952 has been replaced to a large extent by the realisation of the symbiosis in their relationship in the newly-arranged National Forest Policy, 1988. Unless the existing rules and laws, and the ways of working at the government level are subjected to a drastic overhauling to make them consistent with the National Forest Policy, 1988 which provides for the protection of rights and concessions enjoyed by the tribals traditionally, it is doubtful if the provisions will be implemented and desired results be achieved.

5
Product Profile

SOME selected items of EFP are analysed in this chapter in great detail to get a profile covering such topics as processing, marketing and prospects of value addition by which the tribal communities may be more benefited. The items chosen are Kendu leaves, Sal seeds, Gum, Nux vomica, Soapnuts, Essential oils (Palmarosa oil), Lemon grass oil, Siali fibre, Sabai grass and Tasar Silkworm rearing.

Items of essential forest produce

KENDU LEAVES

Kendu leaves which are used for making *bidis* (country cigarettes) are one of the most important EFP of the State. Orissa is noted for the production of Kendu leaves of best quality which are in great demand in the bidi industry. The coppicing of the Kendu leaf bushes all over the State beginning with Basant Panchami Day. This is the most important investment on which quality production depends. After 45 days from the time of bush-cutting, the leaves become ready to be plucked. The collection starts from early April and goes on till the monsoon sets in. Kendu leaves are dried for a week and then kept in Phadi (temporary storage houses). From July to December binding parties move from Phadi to Phadi and do the binding. Collection and processing of Kendu leaves are done by the Forest Department with the funds advanced to the Chief Conservator of Forests (Kendu leaves - KL), Orissa by the Corporation. The entire State is divided into 178 Kendu-leaf units for this purpose.

In the processed area, each lot consists of 720 (one kerris) leaves excluding low cover leaves, one kept above and one below the lot. In the Phala area, a lot consists of 40 leaves (one kerris) excluding two cover leaves. The lot of 20 leaves cost 10 paisa (2L/IP) and the lot of

40 leaves 20 paisa. These lots are packed in bags of 60 kg capacity, each containing 12 bundles of 5 kg each. Thus the Kendu leaf bags in lots 60 qtl. are transported to central godowns where they are handed over to the Corporation for public auction. The Corporation conducts auctions of Kendu leaves at Sambalpur almost every six weeks.

The Corporation has been dealing with the Kendu leaves since 1965-6. In the beginning it started the trade covering some selected areas of the State, but when the trade was nationalised in 1973 the trade with Kendu leaves has increased manyfold.

Annexure VI gives the details of the Kendu leaf trade undertaken by the OFDC from the year 1990 to 1994. This statement giving the Kendu leaf trade shows that in the year 1994 the OFDC had taken delivery of Kendu leaves to the extent of 4.91 lakh qtl.

The turnover for that year records at Rs. 11213.40 lakh, of which Rs. 6552.02 lakh were given to the Chief Conservator of Forest (Kendu leaves) to meet the expenditure incurred for the collection of Kendu leaves. In the same year the OFDC had given Rs. 4869.37 lakh to the State Exchequer in the form of royalty to the government and got its commission to the extent of Rs. 560.67 lakh.

Annexure VI gives division wise procurement of Kendu leaves collection and handing over to the OFDC by the K.L. Department. Besides the increase in revenue, the Kendu leaf trade has been of immense help to the poor and backward people particularly the tribal people. Even in the remotest corners of almost all the districts of the State during the hot summer months when the people are relatively free from their agricultural works, they are kept engaged in such works as bush cutting, Phadi repair, leaf plucking and binding the leaves into lots and transporting them from Phadis to central godowns.

It is a kind of trade which provides labour-intensive employment to millions of people round the year. It is estimated that about eleven million man days are generated during the lean period, that is, from March to June. Even the old, infirm, male and female labourers particularly in western and northern Orissa, find employment in the Kendu leaf trade. At the current rate of plucking, that is, one N.P. per two leaves a person can earn to the extent of Rs. 50 per day whereas the minimum daily wage is fixed at Rs. 25 for unskilled labour. It is estimated that payments to the tune of Rs. 20 crore are made directly to the labourers who are engaged in bush cutting, leaf plucking, drying and binding the leaves in lots, storing them in Phadis and transporting them to the central godowns. In addition to the help given directly to the rural poor, about 50 per cent of the annual net profit according to the State Government is ploughed back for rural development work through the local Gram Panchayats.

There still remains a lot of scope for the improvement of the trade. There are several techniques with the application of which quality production can be achieved. Sincere vigilance over the operational works can ensure the maximisation of production and improvement in quality. On the sales side, attempt should be made to involve the consumers directly in the auction sale so that larger quantities of Kendu leaves are sold to them thereby eliminating the profit which the middlemen get in the business. The real consumers do not attend the auction sale held at Sambalpur because the middlemen who attend the auction sale deliver the Kendu leaves at the doorstep of the consumers. In order to eliminate the dominance of the middlemen in the business it is desirable to take steps so that the bidi manufacturers who are popularly known as 'Seths' come out to take part directly in the auction assuring them of quick service of delivery of the stock at their doorsteps with least botherations.

Unless proper precautions are taken for the storage of Kendu leaves in the godowns, the leaves will be subjected to attack by insects thereby causing wastage of leaves. In order to solve this problem, fragmentation of the central godowns should be necessary to minimise wastage of the leaves.

Another important problem is the smuggling of Kendu leaves. As per the section 9 of Orissa Kendu leaf (Control of Trade) Act, 1961, a grower can keep up to 3.5 qtl. of Kendu leaves without registration in his own house for his consumption. This provision gives scope to the smuggling of Kendu leaves. The smugglers persuade the villagers to store as much as they can, particularly the good quality leaves and when opportunity comes they lift the leaves stored by the villagers by paying them a little higher price than what is fixed by the government.

Another peculiar problem is the Kendu leaf trade in certain areas. What happens is that the pluckers hold back the bigger and good quality leaves with them and sell inferior leaves at government notified centres. The leaves which are held back are used by them for making bidi for some companies active in this area. The companies supply tobacco for rolling the bidis and pay nothing towards the cost of leaves collected by them through their hard labour during the hot summer months.

Round about Bagedia, a village situated at a distance of 50 km from Angul on the road leading to Deogarh, there are many villages inhabited by bidi makers. The Marwaris and Gujratis living in Bagedia and nearby villages Chhendipada, supply tobacco to these people and get bidi from them at the rate of Rs. 27 per 1000 bidis. The people collect Kendu leaves from far-off forests around Rairakhol in two hot summer-months (April-May) and store the collected stock in bundles, each of 3,500 leaves carefully and use them in the coming months. The merchants do not pay them anything for the leaves which they collect against odds.

A CASE STUDY OF BIDI WORKERS OF DASHAHARA SAHI, VILLAGE KHAMAR IN ANGUL DISTRICT

During the months of April and May both men and women go to far-off places like Redhakhol, Bamanda, Chhendipada and Angul which are located at a distance of 20 to 30 km away from the village for the collection of Kendu leaves. Some people of the village have land where Kendu bushes are grown. They prune the bushes every year and pluck the leaves for their own use.

During the months of April and May they collect the total leaf requirements for the entire year. One person requires 48 Podas of leaves (each Poda containing 3000 leaves) for the entire year. Out of one Poda of leaves about 7000 bidis can be made. During the period of collection, the whole family spends their full time in the collection of leaves.

During that period, no bidi making is possible. The collected leaves are arranged in Podas and stored in their own houses. The collection of leaves is in contravention of forest rules because Kendu leaves are a nationalised product but the practice of collection is going on stealthily. Whenever the pluckers are caught red handed the leaves are seized and the entire collection is deposited at the depots, and the pluckers are paid 10 Paisa for 20 leaves.

The bidi workers are supplied with tobacco powder (Masala) by the bidi masters (local middlemen), who also receive the finished product from the bidi makers. The bidi masters receive 4500 bidis from the bidi makers for 1kg tobacco powder. It was told that about 5000 medium-sized bidis can be made out of 1 kg tobacco powder. The bidi masters check the quality of bidis and do the grading as A of superior quality, B of medium quality and C of inferior quality, and accordingly make payment to the bidi makers. This practice has been a substantial source of income to the bidi masters.

In one month, a worker is able to work for 26 to 27 days, and can turn out 800 bidis per day during winter and 1000 to 1200 bidis per day during summer. Thus a worker on an average make one thousand bidis per day. In one month, he can make 26000 bidis. The price of bidis per thousand is fixed at Rs. 27. The monthly income of a bidi maker out of sale proceeds comes to Rs. 702. Thus, the income per year (ten months) comes to Rs. 7000. No cost for Kendu leaves is paid to the bidi maker by the bidi merchants.

The cost of tobacco powder which the bidi master supplies is Rs. 30 to Rs. 35 per kg. The sale price of 35 bidis which makes one Katha, in the market is Rs. 1.80. This shows that the bidi merchant gets a net profit of Rs. 130 out of his investment of 1 kg tobacco whereas the bidi maker gets only his daily wage and no price for the leaves which he collects in the hot summer months.

There are about 150 licensed contractors in the entire Chhendipada block and the bidi trade passes through various channels with loss to the bidi makers and gain to the merchants. The bidi market is located at Gujarat and other places. There are big merchants, who have set middlemen known as dalals. These dalals supply tobacco to the bidi masters who are stationed at the village level to supply tobacco to the bidi makers and receive bidis from the middlemen who live near big villages and towns and have employed supervisors to check the quality of the bidis. At every stage, the bidi makers have to pay something to gratify the bidi masters and the middlemen and in the long run meet a great loss in their business. One possible way by which the bidi workers can be benefited is to organise bidi workers Co-operative Society at village level in a small scale to handle the collection of leaves on a legal basis, storage of leaves and tobacco powder, manufacturing of bidis and marketing of the finished products, directly under the trademark of the society.

SAL SEED COLLECTION

The Sal seeds are collected from the Sal trees which are in great abundance in Orissa. The State has about 25 per cent of the Sal forests of the country and the potential yield works on an average at 15 lakh tonnes. With the progressive dwindling of the forest cover in the State, this potential has been reduced to a great extent. The yield is not uniform in all the years. It fluctuates from year to year. Moreover, the collection has been much below the potential due to various difficulties such as lack of communication facilities in interior areas, non-availability of labour in the interior pockets, lack of storage facilities, short duration of the collection period and adverse weather conditions. The actual collection and disposal of Sal seeds by the Orissa Forest Development Corporation Ltd. for the last 12 years — from 1983 to 1994 — is given in Annexure III(A). It shows that the maximum collection was in the year 1983 and 1989. In the former, the collection was to the tune of 45136 MT and in the latter, it was 41586 MT. But R. Dash in his paper published in the Souvenir brought out by the OFC Ltd. in 1987, has stated that during 1983 which was the bumper crop year, the collection was of the order of 11724.66 MT. This bumper crop in 1983 took place between the years 1978 and 1986. The yield in 1987 rose up to 29,00,000 MT. The yield of Sal seeds in the year 1994-5 was 23,449 MT.

The collection target of the Sal-seed crop for the year 1995 has been fixed by the State Government at 41,000 MT and the TDCC has a target of 5000 MT. The OFDC and the TDCC have been granted lease for the collection and trading of Sal seeds which was nationalised in the year 1983-4. The collection rate of the current crop has been fixed at Rs. 1.75 per kg. The marketing network in respect of Sal seeds is that the Sal seeds collected by a collector is sold to the tribal agents who in turn sell it to the TRIFED.

As the situation stands today, the collector in actual practice meets a single major agency in the field which buys the whole lot from him. This is a type of monopolistic market which has its merits and demerits. There are areas in the State which are inaccessible and situated in remote corners having no transport facilities. The collector does not find market channels in this area and the produce is very often left to rot.

UTILISATION

The oilseeds collected in Orissa are mostly exported to other States. There are no organised industries which are engaged in extraction of oil from oilseeds. Till today there has been no organised linkage between the oil extraction units and collectors. It has now been decided that the oil extraction units which belong to the private sector in the State may be engaged as raw material producers of OFDC and TDCC. The solvent extraction units of the private sector to be linked with the public sector in the state are given in Table 5.1.

Table 5.1: A list of solvent extraction units linked with the public sector

Name of oil extraction unit	*Public sector to which attached*	*Area of operation*
1. Orissa Oil Industries, Sasan	OFDC Ltd.	Sambalpur, Rairakhol, Sundergarh, Deogarh, Bonai
2. Orissa Oil Industries Rairangpur	OFDC Ltd.	Bamra, Baripada, Karanjia, Keonjhar, Puri, Chilika, Dhenkanal, Athagarh, Angul, Athamallik
3. Utkal Oil Ambaguda	OFDC Ltd.	Nowrangpur, Jeypur,
4. Priti Oil Ltd.	OFDC Ltd.	Bolangir, Khariar, Sunabeda,
	TDCC Ltd.	Parlakhemundi, Ghumsar N & S, Rayagada, Boudh Nayagarh, Phulbani, Baliguda, Boudh and Kalahandi.

EMPLOYMENT POTENTIAL

Apart from the utility of Sal seeds, the social gain by the collection of seeds is immense. More than 9 core of Rupees were paid to the Sal seed collectors within a span of about six weeks during 1983 which was the bumper crop year for Sal seeds. The OFDC alone has spent Rs. 6.16 crore towards the collection charges in the year 1993-4 when the yield obtained was to the tune of 35,162 MT.

The Sal seeds are collected in the hot summer months of May and June when there are no agricultural operations and the people are mostly unemployed. Children, old ladies and even disabled persons are engaged in the collection of Sal seeds. Besides the employment potentialities of the collection of seeds, the State pays to the government exchequer in terms of revenue to the tune of Rs. 1.5 crore almost every year.

UTILITY OF SAL SEEDS

Sal seeds were used as a foodstuff by the tribal people. But after the OFDC and other agencies started collecting Sal seeds for commercial purposes, the tribal people have been selling almost all the quantities of Sal seeds they collect from their neighbouring areas and nearby forests for cash.

The Sal seeds contain oil which is extracted by various processes. The defatted Sal cakes are used as cattle and poultry feed. Sal seed oil is suitable for soap making and Sal fat is used for adulterating ghee. The Sal seed kernel yields approximately 13 to 15 per cent hard fat. The seeds are used as raw materials for solvent extraction plants. The solvent extraction is essential for the efficient extraction of Sal fat. This fat is used in the soap industry. The refined bleached Sal fat can be economically used as a substitute of Cocoa butter which is used for chocolates. Sal fat is still being consumed by some tribal people. The analysis of the fat shows that it contains small amounts of vitamin E and traces of vitamin K.

A Sal seed cake containing starch up to 30 per cent is useful in textile industry as sizing material. Since Sal seed cakes contain tannins (14 per cent), this may be used in leather industry. After the starch and the tannins are removed from the cake, it can be used as manure.

Keeping this utilisation of Sal seeds in view, the collection was taken on a commercial basis from 1968. After it was nationalised during the year 1983, the OFDC and TDCC were entrusted with the Sal seed collection and trade. The OFDC has established about 950 collection centres within the area of its jurisdiction, that is 20 forest divisions. The Annexure XIII(A) which gives the picture of collection for the years from 1983 to 1994, shows the rate of Rs. 900 per MT which was fixed for payment to the Sal seed collectors, who are mostly tribals, has increased to Rs. 1750 per MT of collection in 1994.

One of the characteristic features of the Sal trees is that in every 7 years there is one very good seed year, four medium and two poor seed years. The fruits ripen towards May and fall on the ground towards the middle of that month. The people who collect these seeds dry them in the sun and burn the wings or remove them manually and sell the decorticated seeds at the sales centres. 60 kg of kernels are obtained from 100 kg of Sal seeds. The collection continues up to the end of June and it is not allowed to continue the collection beyond June because by doing so the process of regeneration which takes place after the onset of monsoon is affected.

After the seeds are delivered at the purchase centres, these are finally transported to central godowns which are located at railway points or on important roadsides. The disposal of the seeds takes place with approval of a High Power Committee set-up by the government for this purpose. Local industries including solvent extraction units are given preference for the sale of the seeds. Annexure XIII(A) shows that the sale value per MT of Sal seeds stands at Rs. 2721.92 in the year 1994. There was loss in the trade in the year 1992 and 1993; but as shown in the Annexure the year 1994 was likely to be profit year. The Annexure shows that the amount paid to the government in the form of royalty varies from year to year. The royalty paid in the year 1990 rose to as high as Rs. 420.50 lakh and in the following year, that is 1991, there is no record of royalty paid in the year 1994. Annexure III(B) gives figures of expenditure for the collection of Sal seeds.

Annexure III(C) gives divisionwise collection of Sal seeds from the year 1989 to 1994. Taking the quantity collection in MT the divisions can be classified under five categories. Those divisions which fall under category 'A', show a collection which varies from 1500 MT to more than 12,000 MT. Those which fall under category 'E' show a collection which falls below 500 MT. Other categories are distributed between these two highest and lowest categories. The following gives the list of divisions as per their respective categories.

Categories	Divisions
'A'	Jeypore, Nowrangpur, Keonjhar
'B'	Jharsuguda, Rourkela, Karanjia
'C'	Dhenkanal, Deogarh, Jaipur Road, Baripada
'D'	Angul (C), Bhanjanagar, Sambalpur, Rairakhol, Sundergarh.

'E' Bhubaneswar, Angul (KL), Nayagarh, Berhampur, Muniguda, Boudh, Aska, Raygarh, Bolangir, Bhawanipatna.

The Sal seed collection is something which promises to have employment potentialities. It keeps tribal people engaged in collection in the hot summer months when no other employment is available in tribal villages. The Annexure shows that about Rs. 4.12 crore have been paid to the Sal seed collectors within a span of about six weeks in the year 1994 when the collection was to the tune of 22843 MT valued at Rs. 621.77 lakh or 6.22 crore. This is the type of work in which even the old women can take part in the collection without much physical strain. This source of income can help the poor people to tide over the difficulties at the time of natural calamities like drought and failure of paddy crops.

The present production of Sal fat is far below the potential. This is due to the unscientific extraction of fat and harm done to quality of kernels resulting from longer periods of kernel transportation. To overcome this problem portable, decorticators are needed at the field level. A hand-operated portable decorticator which can be taken into interior parts of the Sal forest where power is not available has been devised by the OTRI. There are three such types of Sal seed decorticators which can reduce the volume of 60 per cent from fruit to kernel and also increase the bulk density from 413 to 757 kg, that is an increase of 83 per cent.

Before the Sal seeds are sold, the tribal people undertake the preliminary processing of drying in the sun and decortication at home. No further processing takes place at the middlemen or exporter level. The final processing of oil fat extraction is done by solvent extraction processes at the industries level.

By the traditional method of decortication, the kernel to seed ratio is 50 per cent. Now hand-operated and motorised decorticators are available at cheap rates. The recovery of kernel in the case of mechanised decortication is 66 per cent. Thus the kernel is more and the value addition is to the tune of 33 per cent.

About 24123 MT of Sal seeds were collected in the year 1994-5 by OFDC and TDCC (the potential was placed at 15,00,000 MT having 180,000 MT oil-yielding capacity). This quantity gives 16890 MT of kernels, the fat potential of which comes to 51000 MT at the rate of 300 kg of fat per MT. The use of simple machines makes the decorticating operation quicker. Time is very important factor in the processing of Sal seed kernels. Certain chemical changes take place in the kernels making them less fat yielding and the fat extracted becomes inferior in quality. The fats which are extracted late, are used only for soap making and higher and costlier products. Methods for the preservation of the kernels are yet to be found out, and research is needed in this line.

The tribal people extract fat by their traditional methods. They pound the kernels, boil them in water and collect the fat from the surface of the water. The fat, thus, collected settles down like hard ghee on cooling is used by tribals for edible purposes and also used for lighting. By this process the yield is as low as 3 per cent. The expeller method of extraction by organic solvents has been devised by RRL Hyderabad. But this device cannot be maintained at the individual or small primary society level. However, a group of primary societies can maintain the expeller device. There will be considerable value addition by this method of extraction and an example of value addition is given below :

Calculation of value addition

(Calculated following methods mentioned in the report on MFP, ASC of India, Hyderabad).

(A) Cost of one ton of kernels	Rs. 1750
(B) Price of Sal fat	Rs. 8900 per ton
	28/100 tons of fat
	280 kg of fat
	say, 8900/1000 × 280
	Rs. 2492
(C) Cost of processing	Rs. 675 per ton of fat
	Rs. 189 per 280 kg
(D) Value added	B – (A+C) = Rs. 583 per ton of kernels

The average amount of Rs. 8900, as mentioned above, stands for the internal market price. But the flour for export stands at

Rs. 13,000. If refining is done at the export level, the charges for export including handling of the produce will be around Rs. 1000. At this rate, the value added will be around Rs. 1981 and about Rs. 450 per ton of kernels. Since TRIFED is handling the export marketing of the produce it is expected that with systematic export efforts, the benefits, which will accrue from the export, can be ploughed back to the tribals.

Further and final processing of Sal fat will lead to the production of Sal seed butter, which could replace cocoa butter. The researches undertaken by CFTRI show that Sal seed butter does not have any toxicity and it is fit for human consumption. Hindustan Lever and Cadbury are already using this butter. In this way, at Rs. 17,000 per ton of Sal butter, it can be imagined to what extent there will be value addition to the tribal's income. The various Government Dairy Development Corporations making milk products and having a vast network for marketing present a likely and profitable market for Sal seed butter.

As regards the Sal cake, there are some difficulties regarding its use. The Sal cake seed kernel yields approximately 12 per cent to 15 per cent of hard fat. Thus there remains about 85 per cent deoiled Sal cake-Sal meal. The Sal meal contains about 14 per cent tannin. Work is going on in the ICAC laboratory and in other places to isolate the tannin so that deoiled Sal cake can be used as animal feed. There is a demand of oil cakes in the foreign markets.

1 (a) Sal Fat

Financial Year	*Quantity processed*	*Fat produced (MT)*	*Export of Sal fat*		*Country*
			Qty (MT)	*Value (Rs. in crore)*	
1985-6	93,350	11,975	7,310	17.92	U.K
1986-7	72,690	9,519	5,359	12.51	Italy
1987-8	84,151	10.752	1,558	3.76	Singapore
1988-9	49,362	6,312	1,292	2.96	Malaysia
1989-90	1,47,839	18,641	4,568	10.14	& Japan
1990-1	53,668	7,031	2,312	6.16	
1991-2	58,523	7,489	3,312	13.99	
1992-3	74,032	9,598	5,995	21.45	
1993-4	1,03,747	13,467	1,914	7.42	
1994-5	45,635	5,964	2,388	9.96	

(As from S.K. Pal's paper Export Potential Table 4 - NOVOD Background Papers)

(b) Sal seed extraction

Financial Year	*Production of Sal seed extraction*	*Export of Sal Fat*		*Country*
		Qty (MT)	*Value (MT)*	*(Rs. in crore)*
1985-6	78,900	14,500	0.36	Romania
1986-7	61,864	19,485	0.97	Holland
1987-8	71.412	23,884	1.44	U.K, Germany
1988-9	42,326	21,517	1.73	Italy
1989-90	1,27,245	74,547	6.06	Singapore
1990-1	46,256	43,377	4.77	Hong-Kong
1991-2	50,501	34,830	6.93	Indonesia
1992-3	63,394	55,189	10.47	
1993-4	89,655	37,388	5.62	
1994-5	39,307	35,855	6.86	

(As from S.K. Pal's Paper Export Potential Table 5 - NOVOD Background Papers)

Sal seeds are produced in Orissa in abundance. These seeds are of great economic value. The tribal people in particular can derive substantial benefit from this produce which is available in their own habitat. They can use the seed and its oil as edible items. Sal cakes or Sal meal without the tannin content can be popularised as animal feed, both for cattle and poultry among the tribal people. The Sal seed is a source of manure which can be applied in the fields. Considering the various utilities of this product, necessary steps need to be taken for a greater yield and procurement and systematic use and marketing of the product for greater value addition to the tribal communities. Some of such steps needed to be taken, are listed below:

1. As the collection period is of about six weeks in hot summer months and the inaccessibility of many areas having thick Sal forests have stood in the way of collection efforts whereby the present collection level is at a very low level.
2. Mechanical devices should be used for speedy and efficient decortication.
3. Solvent extraction plants may be set-up at places in clusters of primary societies of tribal communities. The transportation facilities may be provided, so that less time is required for transporting the kernels to the processing units.

4. Small industries may be set-up in tribal areas employing the tribal people in soap making from Sal fat. Organisational details and accounting procedures for keeping records of benefits accruing to the tribal people from such tribal industrial units will have to be worked out. Motivation is required to spread education among the tribal people enabling them to keep such accounts by themselves and run the industry smoothly.

5. Development of an internal market for Sal seed butter should be handled by the TRIFED in a systematic manner. OTRI has patented an improved method of refining and bleaching dark-coloured oil into a light-coloured one which can be used as Cocoa Butter Substitute (CBS).

6. The farmers should be mobilised and educated to use the deoiled Sal cakes which are accumulating at various solvent extraction plants unused. Concerned research units should take up work concerning isolation and reduction of tannin content leading to the improvement of the feeding quality of the deoiled cakes.

7. Sal resin yields an essential oil called Chua used as fixative and for flavouring chewing and smoking tobacco. It is used for ritual purposes and in healing of ear trouble and cutaneous diseases. It is necessary that the tribals should be trained in proper tapping and collection of the produce without least injury to the trees.

Export of various oils and fats and their extraction have been derived from NTOs by Solvent Extractions.

Gum

There are two types of Gum, Gum Arabic and Gum Tragacanth. The former is obtained from certain species of the genus Acacia. The gum which exudes from *Acacia nilotica* (L.) Willd. ex Del. (Babul) may come closer to the real Gum Arabic collected from *Acacia senegal* Willd. in Sudan. The gum obtained from *Anogeissus latifolia* (Roxb. ex DC.) Wall. ex Guill & Perr. (Dhaura) called Ghatti gum may serve as another substitute for Gum Arabic.

The Gum Karaya otherwise known as Kartira Gum or locally known as Genduli Gum is obtained from *Sterculia urens* Roxb.

(Genduli). It is of the Tragacanth group. Out of 7000 tons of annual production in India, half of it is obtained from Madhya Pradesh. Orissa produces only 100 qtl. of Genduli gum.

Gum Karaya needs processing to make it usable. But the technical know-how is lacking in the country. The tribal people who generally collect the gum, have no idea about the proper collection procedure, cleaning of impurities (bark, mud, sand, etc.) and grading and packing. Indiscriminate tapping ruins the plants. Merchants sometimes instigate the tribals to tap more times at short intervals and the plants are felled and lifted away which causes depletion of this plant species.

It is necessary that research and training should be undertaken on proper methods of collection, grading, packing, processing and replenishment so that quality product is obtained adding value at the tribal level. Primary procurement centres need to be set-up and also small grading centres may be established within a cluster of procurement centre. Some small machines such as gravity separator, colour separator, air separator, crusher, etc., may be needed in processing so that the grading could conform to the standards of foreign importing agencies. Nevertheless the tribal manual skill and indigenous knowledge may be explored and the tribal expertise and labour utilised for their benefit.

The export trade is now handled by many middlemen and traders. Therefore pilferage and malpractices happen. The exporters compete with one another to get the export contract and the importers exploit this situation. In view of this, it is desirable that a single agency like TRIFED should handle the export which will ultimately benefit the tribals by adding value at their level including their employment.

Gum is not a nationalised item in Orissa. Therefore there is no systematic collection of the product. Various types of gum such as Genduli, Char, Bahada and Dhaura gum are collected. But they are mixed together. Proper statistics about the quantity procured are lacking. There is thus large scope for organisational work. Further research and training are needed to develop suitable substitutes for imported Arabic gum, and evolve proper devices for collection, storage and trading.

A cost benefit analysis is given in the report on MFP, 1977-8, Administrative Staff College at Hyderabad. It is seen that value added by grading the gum comes to Rs. 1121 against the price of graded gum per ton which comes to Rs. 4835. Orissa procures only 100 qtl. of gum and if it is graded it becomes much less. It can locally be sold having no prospect of export.

Lac (Lac insect Lacifer lacca)

With the development of the trade, lac is cultivated in many places. Ranchi in Bihar is one of the important lac cultivation centres. The method of propagation is like this.

Twigs of a tree used in the previous year having deposits of crowded larvae are tied to the twigs of a host tree to be brought under lac cultivation. Such lac sticks having mature female insects are called brood lac. The trees most suitable for lac are *Butea monosperma* (Lann.) Taub. (Palasa) and *Schleichera oleosa* (Lour.) Oken (Kusum). The insects are gradually hatched from the larvae and crawl out to feed on the sap of the host tree. In this process, a thick resinous fluid, which exudes from the body of the insects, forms a hard incrustation of lac over the whole twig. The lac from the Kusum tree is of the best quality fetching the best price. There are two crops of lac annually, one in the hot weather and the other in winter.

Lac cultivation requires intensive operations such as pruning, harvesting, inoculation, etc. The brood should be cut at the proper time, otherwise success cannot be achieved to the fullest extent in lac cultivation. The brood lac must be utilised within a few days after harvesting, otherwise most of the larvae will be destroyed.

There are two forms of lac, namely Ari and Phunki. In the former case, it is cut from the host plant and in the latter case, it is collected from the brood lac. When the lac has been removed from the sticks, it is known as stick lac.

The stick lac is dried, crushed, and winnowed to free the substances from the bits of wood, mud, sand, etc. Storage of lac for a long time causes deterioration. The substance requires washing to remove the colouring matter, insect bodies and other impurities sticking to the substance. Washing can be done either manually or mechanically in drums with a measured quantity of water. The washed lac is graded according to the quality of the substance. This product is called seed lac. The cost of processing stick lac into seed lac is estimated to be around Rs. 250 per ton. This preliminary processing at the level of the tribal cultivator adds to the price obtained by them.

Seed lac is used for conversion into shellac. The method of shellac conversion is by heating it up. The cost of processing into shellac is estimated at Rs. 1000 per ton. The yield is roughly around 55 per cent by weight of stick lac. Kusumi seed lac yields more shellac and aged seed lac yields less than fresh seed lac. Solvent extraction process gives a higher and richer quality yield than the heat process method.

Orissa was producing Kusumi lac to the extent of 1381 MT in 1947-8 and again in 1967-8. But it has declined to only 25 MT at present. The study shows that the TDCC has procured Kusumi lac only from two districts, namely Keonjhar and Sundergarh and the quantity procured was to the extent below 300 qtl. It is stated that the production of Kusumi lac in Orissa has suffered mainly due to non-availability of brood lac. The brood lac farms which were maintained by the Forest Department, are not existing now. The TDCC gets the brood lac from Ranchi Research Centre, but the mortality of the larvae is so great that only 20 per cent of success is achieved in propagation. The brood lac supplied by some private agencies succeeds to the extent of 80 per cent in propagation. The study of MFP, sponsored by the Ministry of Home Affairs, has worked out value addition in respect of lac cultivation. The calculation is given in the report on Minor Forest Produce, Administrative Staff College, Hyderabad, India at p. 137-8. It shows that if processed at home, there will be value addition to the extent of Rs. 1138 per MT of seed lac of Rs. 3250. These are back-dated figures which do not apply to the present price level.

Considering various situational factors with respect to lac cultivation, marketing network, trading, price, storage and propagation, it is necessary to stabilise the market and prevent fluctuations in demand. Moreover it is necessary to introduce complete state trading of the commodity to do away with speculative and exploitative practices and thereby ensuring a legitimate financial benefit. To achieve those objectives it is necessary to take the following steps :

1. Lac cultivators co-operative societies may be formed among the tribal people. The members need to be trained in scientific lac cultivation, starting from pruning, harvesting, inoculation, phunki removal, supply of brood lac and various other seasonal operations.

2. Demonstration plots have to be set up, and R&D are to be taken up to assess the result, and popularise result demonstration. All the activities have to be co-ordinated so that the governmental assistance is channelled through the lac cultivators co-operatives and take delivery of the seed lac and arrange for its transportation and storage.
3. The TDCC will pay a predetermined support price for the seed lac to the lac cultivators co-operatives and take delivery of the seed lac and arrange for its transportation and storage.
4. Seed lac may partly be sold inside the country, partly exported to other countries, and partly processed to shellac. In the case of internal marketing, the State Lac Marketing Federation may receive the seed lac from the TDCC and arrange for domestic marketing. The TRIFED may take up the role of a State Lac Marketing Federation because they have the requisite advantage of being able to develop marketing strategies, stabilise uniform prices and avoid competition.
5. The TDCC itself will be entrusted with the task of processing the seed lac into shellac, and the finished product will be handed over to TRIFED for marketing.
6. Domestic marketing of lac is very important for stabilising the lac trade. At present, only about 10 per cent of the lac produced in the country is consumed inside the country. In such a situation, the lac cultivators have been the victims of a fluctuating export trade.
7. Lac was used for making gramophone records. Its largest consuming channel is now computer manufacturing, surface coating both for decorative and insulating varnishes, lacquers of various kinds, woods finishing, metal coating, playing cards, linoleum, oil cloth, etc.
8. Shellac is used in the form of insulating varnishes, mica, paper insulation, spark shields, etc. It is also used as primary raw material in leather dressing, straw-hat stiffening, sealing wax, adhesives and cements, glazed paper, nail polishes, dental plates, jewellery fittings, grinding wheels. Considering the multiplicity of use, it is necessary to have a good marketing organisation to serve the demand for giving feedback to the

product development agencies like Lac Research Institutes for setting up a network and for adopting vigorous marketing techniques.

9. Average shellac contains 68 per cent resin, 10 per cent lac dye and 6 per cent waxy substances. The Lac Research Institute should work to find out the new uses of these component parts in the industrial field. At the same time, the Institute must have feedbacks from the marketing agencies regarding commercial potentials and trade specifications concerning the use of shellac and its constituent parts.

Mango kernels

Mango kernels is procured by the Utkal Forest Products Ltd. In 1991-2, collection was to the tune of 77,478 MT. But in 1992-3, it came down to as low as 165,848 MT. Last year, the quantity of mango kernel touched the mark of 1978 MT. The TRIFED, which received mango kernel at the rate of Rs. 3250 per MT, got 190 MT of kernel processed at Orissa Oil Industry and sold the extracted oil of 140 MT to Hindustan Lever at the rate of Rs. 23,000 to Rs. 26,000 per MT. The total quantity of DOC by product of 690 MT was transferred to Kakinada branch of TRIFED.

The commercial feasibility of mango kernels (61-83 per cent of stones) as a source of vegetable oil (oil content, 6-14 per cent) was demonstrated for the first time in 1974 by a private firm from Maharashtra. The OTRI and G.B. Pant University of Agricultural Technology developed hand and power operated mango stone decorticators for the separation of kernels from the stones. Drying and storage of kernels, extraction, refining and utilisation of oil and deoiled meal, etc., are being studied at several institutions. Mango kernel oil is a hard fat (iodine value 32-60.7) with oleic (43.6-50.3 per cent) and stearic (36-43.1 per cent) acids. The fat comprises mainly of disaturated (30-45 per cent) triglycerides and by virtue of this glyceride composition, it is in demand as cocoa butter substitute, and is exported. The use of fat is also allowed in vanaspati (10 per cent). High FFA fat is used in soap making. However, the production of mango kernel oil has not exceeded the 2500 tonnes mark because of the problems in organising the collection of mango kernels. Further, at the time of collection, moisture content in stones is about 25-30 per cent. If not decorticated and dried immediately, the FFA content increases rapidly resulting in high processing losses during the refining of the kernel oil, and also affecting the quality of oil for edible and other high-priced usage.

Calculation of profit

(As per the method adopted in the report referred to above of ASC of India, Hyderabad)

(a)	Dried mango kernel (1 ton mango seed = 600 kg of kernel)	Rs. 1665 per MT
(b)	Cost of extraction	Rs. 1340 per MT of kernel
(c)	Fat 8 to 9 per cent Cost of refining @ Rs. 2000 per ton loss in refining	90 kg of fat per ton of kernel Rs. 180 25 per cent = 67.5 kg refined fat
(d)	Price of refined fat for 67.5 kg Price of fat	Rs. 1755 Rs. 1755 out of 1 ton of kernel
(e)	Deoiled meal 86 per cent Price of 860 kg	860 kgs Rs. 3000
(f)	Total price of fat and meal	Rs. 4755
(g)	Total cost (a+b+c)	Rs. (1665+1340+180) = Rs. 3185
(h)	Profit (f–g)	Rs. (4755–Rs. 3185) = Rs. 1570

Thus it is found that extraction of oil from mango kernels is a profitable business. But the extraction of oil by solvent method should be taken up at the tribal level. The tribal people can be employed in the oil mills to learn the process of extraction and at the same time earn from this occupation.

A single agency like Utkal Forest Products Ltd. has been given lease to collect mango kernels from the tribal people. The primary gatherers have no choice but to dispose of the quantity of their collection to this private agency who is obviously more interested in getting more profit at the cost of tribal welfare. In view of this, the TDCC is the most suitable agency to procure the mango kernels.

Although the government corporations like TDCC is not likely to be exploited, the common weakness of this agency is that its field units are not adequately staffed and the funds, placed at their disposal for payment to the primary gatherers, are not adequate. Due to lack of incentives, they do not touch the interior and remote areas, and therefore the primary gatherers sell their collections at a throw-away price to any private person visiting them at their doorstep.

This system is also be set with certain disadvantages and these drawbacks have suffered with the introduction of Joint Forest Management. Under this management, it is stipulated that the tribal people are required to sell the produce only to the official lessee and they cannot dispose of the collection from the forest to anybody of their choice on the basis of the highest price offer. This dampens their motivation to protect and manage the forest. In view of this it is better that both private agencies and government corporations be allowed to operate together in a competitive setting.

Sabai Grass (*Eulaliopsis binata* (Retz.) Hubbard)

Sabai grass grows wild in the forest and in uplands. It is being cultivated extensively in many places. Upland terrain, good drainage and low rainfall are most favourable for the growth of Sabai grass. It is harvested in the months of August and September. A quintal of grass at the time of harvest fetches only Rs. 175 to Rs. 200. But the rate shoots up to Rs. 375 to Rs. 400 at the time of the off-season.

The grass is used for making ropes which are used for making furniture, carpets, foot mats, table mats, sofa sets and many different items of decorative objects of are displayed in walls and other places of houses.

One hectare of land gives in the minimum an yield of 110 qtl. of grass. The yield, which is quoted here is obtained by traditional methods of cultivation. If it is grown by scientific methods, the yield will increase by 10 times and the grass in the plot of land can be harvested twice annually for a period of 10 years without application of any inputs during this period. A cost-benefit analysis of Sabai grass cultivation is given below.

Cost of inputs for increased yield

The cost of inputs for the first year, and the harvesting cost and yield rate for the subsequent 4 years are given below. The calculation is based on one hectare unit of land.

Based on the above-mentioned figures a calculation of Sabai grass cultivation in 30 ha of land is given below for the sake of illustration. After the lands are reclaimed and necessary inputs are applied, these 30 ha will produce 3000 qtl. of grass at the rate of

100 qtl. per hectare (taking the 4th year as the base), which will fetch a net income of Rs. 42,000 annually for these 30 ha.

If the entire amount of grass is used for making ropes, there will be more profit. 3000 qtl. of grass will give 2330 qtl. of ropes at the rate of 85 kg ropes per quintal of grass and these ropes will fetch in the minimum Rs. 8,92,500 during the time of harvesting of grass at the rate of Rs. 350 per qtl. But during the months of February to July when the grass is scarce, the price of these ropes will rise to Rs. 19,12,500 @ Rs. 750 per qtl. If the cultivators or a co-operative society organised by Sabai grass growers can store the grass and wait two or three months after the harvest, then they can get on an average a price not less than Rs. 14,00,000.

The advantages of Sabai grass growing are many; one time investment will enable the grower to harvest the produce from a single plot of land for a period of 10 years. It proves to be a very good income-generating scheme based on local human and natural resources and raising the economic standard of people. Other positive points of the scheme are that the gestation period is only one year. It requires only one time investment, and yields increase almost 10 times of the present level. Of all the plus points, the most crucial one is the sustainability which the programme achieves and the self-reliance which is fostered among the beneficiaries to keep the activity going with an increasing rate of economic benefits.

Cost of inputs of increased yield

The cost of inputs for the first year and the harvesting cost and yield rate for the subsequent four years are given below. The calculation is based on one hectare unit of land.

First year	Amount (Rs.)
Preparation of land	
(a) Ploughing (22 ploughs × Rs. 70 per plough)	1,540.00
(b) Cowdung (13 cartloads × Rs. 100 per cartload)	1,300.00
(c) Digging pits (37 man days × Rs. 25 per man day) (1 meter spacing)	925.00
(d) Filling pits with cowdung (37 man days)	925.00
(e) Applying pesticides (37 man days + cost of pesticide)	1,000.00
(f) Sabai seedlings (2.5 trucks)	4,000.00

(g) Planting in July (50 man days)	1,250.00
(h) Fencing	4,000.00
Total	***14,940.00***

Second Year	Amount (Rs.)
Harvesting	
1st crop (August - 25 qtl. of grass × Rs. 175 per qtl.)	4,375.00
2nd crop (November - 10 qtl. × Rs. 200 per qtl.)	2,000.00
Total	***6,375.00***
Harvesting cost	
1st crop (50 man days)	1,250.00
2nd crop (20 man days)	500.00
Total	***1,750.00***
Net Income	4,625.00

Third Year	Amount (Rs.)
Harvesting	
Ist crop (50 qtl. of grass × Rs. 175 per qtl.)	8,750.00
2nd crop (20 qtl. × Rs. 225)	4,500.00
Total	***13,250.00***
Harvesting Cost	
1st crop and 2nd crop (140 man days)	3,500.00
Net income	9,750.00

Fourth Year	
Harvesting	
1st crop (70 qtl. × Rs. 175)	12,250.00
2nd crop (30 qtl. × Rs. 225)	6,750.00
Total	***19,000.00***
Harvesting Cost	
Ist crop and 2nd crop (200 man days)	5,000.00
Net income	14,000.00

Fifth Year

Harvesting

1st crop (70 qtl. × Rs. 175)	12,250.00
2nd crop (40 qtl. × Rs. 225)	9,000.00
Total	***21,250.00***
Harvesting crop	
1st crop and 2nd crop (220 man days × Rs. 25)	5,500.00
Net income	***15,750.00***

The yield obtained in the 5th year is maintained till the end of the 10th year with no further inputs.

The NABARD has worked out the economics of Sabai grass cultivation in 1 acre of land which is given below.

Financial statement

	Year	*1*	*2*	*3*	*4*	*5*
A.	**Cost in Rs.**					
	Cost of plantation	1825	500	500	500	500
	Cost of twisting wheel	300				
	Rope output/year (kg)	2700	2700	2700	2700	2700
	Grass utilised (kg)	3105	3105	3105	3105	3105
	Grass yield from own source (kg)		400	900	1200	1400
	Grass purchased from market (kg)	3105	2705	2205	1905	1705
	Cost of grass purchased (Rs. 3/kg)	9315	8115	6615	5715	5115
	Labour cost (of family @ Rs. 30/ day)	8100	8100	8100	8100	8100
	Total Cost	19540	16715	15215	14315	13715
B.	**Income from rope @ Rs. 8 kg**	21600	21600	21600	21600	21600
C.	**Surplus Income**	2060	4885	6385	7285	7885
D.	**Repayment Schedule**					
	Loan O/S (being the project cost)	4800	4800	3600	2000	

	Loan instalment		1200	1600	2000
	Interest from loan (Rs. 5100) @ 12 per cent	576	576	432	240
	Total repayment	576	1776	2032	2240
E.	**Surplus with farmer**	1484	3109	4353	5045

The unit cost can be modified if more land can be brought under cultivation. A plantation of 1 ha can meet the grass requirement of a family (3 members) from the 4th year onwards.

1.	Cost of raising 1 acre of Sabai grass (1yr.)	Rs. 1825.00
2.	Cost of twisting wheel	Rs. 300.00
3.	Cost of grass required for 2 months @ Rs. 3 kg for 600 kg	Rs. 1800.00
4.	Family labour for 1 month @ Rs. 30/ day	Rs. 900.00
	Total	**Rs. 4825.00**
	Say	**Rs. 4800.00**

The first calculation based on 1 ha of land shows a net income of Rs. 14000 whereas the corresponding figure for 1 acre of land in NABARD calculation shows Rs. 5045. The first income is a little higher, because the farmer has done cultivation without any loan component by which interest was not needed to be calculated. Moreover no grass was purchased as it was also not needed.

Mahua Seeds (*Madhuca indica* Gmel.)

Mahua trees are extensively found in different parts of Orissa. Its flowers are eaten and used for brewing liquor (Mahuli wine). Mahua seeds are oil yielding which are edible. One quintal of Mahua seeds yields 70 kg of kernels, the oil content of which is about 50 per cent.

The cake containing 10-14 per cent oil is subjected to solvent extraction. Mahua oil, a semisolid hard fat is widely used in soap making. The refined oil is edible according to PFA rules. Crude oil is produced and consumed locally by the tribal people in Orissa. Mahua oil is also used in the manufacture of vanaspati, linoleum, greases, and its fat is used for making candles. The fat finds its use for medicinal purposes, particularly in skin diseases and rheumatism. As its contains sizeable proportions of symmetrical disaturated and monosaturated glycerides in the fat, it can be used for production of cocoa butter and substitute fats. Since processing is highly expensive it is not much in demand.

Mahua cake and deoiled meal is used as manure, as wormicide in turfs, and as fish poison because of the toxic effects of Saponin in the meal. There is a possibility that through biodegradation, the manurial value of the cake can be improved and bio-gas can also be obtained. The detoxified cake can be utilised as poultry and cattle feed. However, further studies are required to establish its use as cattle-feed and for other commercial application.

Mahua fruits ripen in the months of May and June, and therefore seeds are available mostly in the summer months.

Not all seeds find their outlet outside the village for marketing. Whatever quantity of seeds are retained, are used for extraction of oil by indigenous methods or by village Ghanis. The seeds which go out of the village, reach the market through a chain of middlemen, and ultimately reach the organised sector of soap-making. 42-45 per cent yield of oil is extracted from the seeds by solvent extraction and 30-35 per cent yield of oil by the expeller method. The yield depends upon the quality of seeds. Storage of these seeds for a longer duration diminishes the yield whereas the fresh seeds give a higher yield. Considering these points, the tribal people dispose of the Mahua seeds in the months of May and June when the produce fetches a higher price. Considering the above-mentioned points, the following steps have to be taken to assure benefits to the tribals in Mahua-seed trading:

1. The Forest Department will take up the growing of Mahua trees in groves instead of allowing them to grow wild and in a haphazard manner. A Mahua tree on an average lives for about 60 years and it bears fruits for 30 to 40 years. The annual collection is estimated to be about 40 per cent. This is because of the fact that the collection season is very short and some of the trees are in inaccessible areas. In the year 1993-4, the quantity collected was 761.861 MT, which is much below the potential.

2. The quantity of seeds collected in proportion to the total crop is much less in forest areas than near villages. Hence, it is necessary to do something for organised harvesting and collection in the forest areas.

3. Extraction of fat at the local level should be emphasised. The tribals should be motivated to meet their oil need from their own sources rather than depend upon edible oil supply coming from the organised sector.

4. Village Ghanis which are in the process of disappearance in the village should be revitalised. These should be set-up at primary collection centres for extraction of oil. In fact, the extraction at the village level will be around 38 per cent and relatively pure oil will be obtained. The fat catering to the local needs can be used as the input to soap-making and candle-making industries which can be set-up by a cluster of primary societies covering a sizeable group of villages. This arrangement will have the following advantages:

   (i) Assuring greater benefit to the collectors.

   (ii) Eliminating dependence of tribal people on costly products.

   (iii) Preventing the presence of solvents in the Mahua deoiled meal which is the case when oil is extracted by solvent processing method.

   (iv) Enabling oil extraction from fresh kernels, getting quality oil of larger quantity and avoiding storage and expensive transport charges.

Myrobolans (*Terminalia chebula* Retz.)

This is largely a tannin-yielding EFP. The fruits are also used for medicinal purposes. The tannin content in myrobolans varies from 12-49 per cent with the average coming to 30 per cent. Myrobolans are used for water softening by the railways, as a mud thinner in oil-drilling operations and also in the manufacture of ink and as a mordant in the drying of cotton and for weighing of black silk. By far the most important of the myrobolans is the fruit of *Terminalia chebula* Retz. (Harida) known as *Chebulic myrobolan*. Two other myrobolans are also known among the items of EFP. These are *Terminalia bellirica* (Gaertn.) Roxb. (Bahada) and *Emblica officinalis* Gaertn. (Amla). Commercially the former is known as *Belleric myrobolan* and the later as *Amblic myrobolan*.

Myrobolans are most widespread in Orissa. Annually 13120.8 qtl. of myrobolans are collected from different parts of Orissa. The areas where myrobolans are available in larger quantities, are Koraput, Kalahandi, Mayurbhanj, Sundergarh, Angul, Bolangir, Nowarangpur. Only 40 per cent-50 per cent of the crop is collected from the forests. Large amounts of myrobolans are utilised in the country whereas a small amount is exported in various forms to foreign countries. The agencies like the OFDC, TDCC, and AMCS Tikabali are lease holders for the collection of the myrobolans.

The fruits after collection are spread on the ground and allowed to dry in about two to three weeks. After the fruits are dried, they become ridged and the ridged fruits are considered the superior quality and best for tannin. The sun-dried myrobolans can be stored for two to four years without causing any harm to the fruits.

The stones of the myrobolans constitute 20-50 per cent by weight of the fruits and contain only about 2-4 per cent of tannin. The trade of myrobolans is confined to crushed seeds. Crushing is done by hand-pounding, the ratio of crushed nuts to whole nuts on an average is 3:5.

In addition to whole fruits and crushed myrobolans, solid extract of myrobolans has been a popular and regular article of marketing. The extract is prepared in factories and despatched in the form of solid blocks, containing 50-60 per cent of tannin. The different stages of processing before the material is marketed are as follows:

1. Storing and drying
2. Crushing
3. Extracting.

Most of the tanneries prefer to make their own liquor mixed with other tanning materials. Therefore they prefer crushed myrobolans to extract tannin. The general picture of the myrobolan consumption is as follows :

1. As crushed myrobalan	70-75 per cent
2. As whole nut	20-25 per cent
3. As myrobolan extract	5 per cent

Thus the marketing of myrobolans the tribal communities should concentrate on crushed and uncrushed myrobolans.

There is a vast potential for the use of tannin material in the rural sector. The production of tanned hides in the country is estimated to be at 30 million pieces and about 40 per cent of the tanned hide are processed in the cottage and village sector. 90 per cent of the production of footwear is from the manufacturing units in the medium and small cottage industries sectors.

For assuring greater benefit to the tribal people who collect the myrobolans, attention should be given to the following points:

1. Special arrangements should be made for the collection and procurement of myrobolans by the Government of Orissa undertakings to the fullest extent.
2. Necessary facilities may be provided and the tribals educated in improved methods of sorting and drying, resulting in better yields. These arrangements will have to be made at the procurement centre or at the collection points.
3. Toothed-roller crushers should be installed at the cluster of procurement centres for crushing the nuts. The fruits should be cleaned and sieved before crushing. These steps will automatically reduce the volume and make the transportation easier, quicker and more efficient.
4. The tribal communities in general are much averse to work on anything related to hide and leather. In view of these, tanneries may be set-up in rural areas inhabited by those communities who handle such products without any social stigma. The tribal communities who collect the myrobolans may have links with the tanneries so that the raw materials are supplied by the collectors to these tannin units. It is necessary to work out a few feasibility schemes for establishing tanneries in the areas of the State.

It is the custom in rural and tribal areas that the carcasses are removed from the villages by the Doms, a scheduled caste community of Orissa. They also clean the hides and skins and supply them to tanneries. The trade in hides and skins is in the private hands of non-tribal communities and the tribal people have nothing to do with this trade. But the collectors of myrobalans are mostly tribal people, hence it is necessary to establish a link between the collection of myrobolans and consumer organisations. A systematic organisation of the supply and consumption of the produce is sure to enhance not only the economic benefit to the tribal people but also the village tanneries.

A major portion of the output of the tanneries lies in footwear making. The major raw materials come from the tribal areas, but the tribal people in a large majority are not beneficiaries of the finished products. It is necessary that the KVIC and any other such organisations have to make schemes for setting up small footwear units adjoining to the village tanneries, or as separate units in which the tribal communities can be employed in jobs which should be in conformity to their cultural customs and traditions. Apart from footwear, several kinds of leather goods, sports goods can be manufactured. These items can be handled by the tribal communities without any social stigma.

Collection of myrobolans, hides and skins from the villages can be operated by the primary procurement societies organised among the tribal beneficiaries. Similarly the tanneries and production of leather goods can be operated by a group of primary procurement societies. If it is possible, a few myrobolan extraction plants may be set up and organised, and operated by TDCC or by an independent organisation under the control of the State Government department responsible for development of cottage and village industries. The profit from these units should be ploughed back to the tribals through the tribal sub-plan partly for tribal development and partly as cash remuneration. However, the details about plough back of benefits, and the quantum and modalities, are matters of policy-decision lying with the government.

Nux vomica

Nux vomica is one of the well-known EFP, which is collected from different parts of the State. The last years figure shows that on an average, 2130 qtl. of *Nux vomica* were collected. The agencies, which are authorised to collect this produce are TDCC and OFDC. The OFDC collects this produce from 5 regions namely Bhubaneswar, Dhenkanal, Angul, Baripada and Karanjia, whereas the TDCC collects it from other parts of Orissa. The collection season starts in December and goes up to the end of February.

The toxic and medicinal properties of *Nux vomica* are composed of two alkaloids namely Strychnine and Brucine. Both are poisonous and are extracted from the seeds although other parts of the plant contain them in lesser concentration.

The fruits, which look like small orange fruits, are bitter and the seeds are collected by removing the pulp. The seeds collected from the ground are inferior in quality to the seeds collected from inside the fruit by removing the pulp. The seeds are washed thoroughly, dried and sorted out before marketing. Both the seeds and pharmaceutical products are exported to USA in large quantities.

The processing of the seeds includes pulverisation through expellers and extractors. Strychnine, Brucine and other alkaloids are used in preparation of drugs, which are used to stimulate appetite and for treatment of nervous disorders and paralysis. The procurement rate per quintal of *Nux vomica* comes around Rs. 700 and is sold at Rs. 1600. This produce is collected regularly for the last 10 to 15 years.

Extraction of alkaloids is a complicated process which can be done only in factories. The *Nux vomica* seeds are pulverised and then subjected to expression and extraction. The alkaloids are isolated by chemical methods and purified by distillation.

One ton of *Nux vomica* yields up to 2 per cent alkaloids of which 40-42 per cent is constituted by Brucine and 35 per cent by Strychnine. It means that 1 ton of *Nux vomica* gives 85 kg of Brucine and 7 kg of Strychnine. The Regional Research Laboratory at Bhubaneswar has done work on the extraction of Strychnine and Brucine and also worked out production economics of the extraction process to improve profitability.

Sometimes back, the situation was that there was no market for the alkaloid extracts of *Nux vomica* inside the country because of lack of technological development to safeguard against the side effects of usages of Strychnine and Brucine. In the export market also, the medical use of alkaloids is very much limited. Most of these poisonous alkaloids are utilised outside the country for the manufacture of rodenticides. In the case of Brucine, disinfectant sprays are also manufactured from these poisonous products. There is a good deal of opposition from various organisation particularly the Wildlife Preservation Society against use of Strychnine in the manufacture of rodenticides. This product does not promise a good market.

In view of this, it is necessary to undertake research and development of this product in such institutes as Central Drug Research Institute and the Central Council for Research in Indian Medicine and

Homoeopathy. Any success in research undertaking may give scope to work out schemes for greater benefits to the tribals.

Soapnuts (*Sapindus emarginata* Vahl.)

Soapnuts are one of the important items of EFP which are available in forests throughout Orissa. TDCC of the State has got the lease to collect the item. Last year, about 148 quintal of soapnuts were procured by the TDCC.

Soapnuts are used for cleaning silken clothes. Besides its cleaning properties, the soapnuts contain saponin which has various uses. The saponins are used in detergents, shampoos, textiles, food, pharmaceuticals and the photo-film industry. The entire quantity of saponin requirements in the country is presently being imported from other countries.

A process has been developed for isolation and purification of saponins from soapnuts by the Regional Research Laboratory, Bhubaneswar. Annexure XIV gives details of properties and processing of soapnuts.

Fibres

Siali fibres (*Bauhinia vahlii* Wt. & Arn.)

The Orissa forests are very rich in various types of fibres, climbers and creepers. Most of these products are used in making ropes and baskets. One of such fibres is Siali fibre which is found in abundance in the forests. The Birhors, one of the primitive tribes of Orissa, collect the fibres from this plant and depend upon it for their livelihood. They are a semi-nomadic tribe moving from one district to another, and they live in places which are very close to the forest as well as weekly markets and villages inhabited mostly by cultivating castes.

The items which the Birhors make out of Siali fibres, are slings (Sika), halters for cattle (Pagha), ropes for drawing water (Barehi), halters for buffalos (Kadadogha), goats (Meromjoda), and yoke ropes (Dhaunri). The fibres are also used for making certain articles like slings which are used for decorative purposes. The items which the Birhors produce out of Siali fibres are in great demand among the cultivators and cattle breeders. They make baskets out of Siali creepers which are used by other tribal communities for pressing oil from oil-yielding seeds.

One load of Siali bark is sufficient to turn out ropes and baskets worth Rs. 100 and a person can exhaust a single load of barks in two days. On an average, a Birhor turns out finished products of bark worth Rs. 50 per day. Both men and women are skilled in making fibre products.

Essential oils

Palmarosa oil

This oil is obtained from *Cymbopogon martinii* (Roxb.) Wats (Rosha grass). Morphologically it has two varieties Motis and Sofia. The former is known as Palmarosa and the latter as Ginger grass. Palmarosa is the best natural source of high-grade geraniol with a content up to 95 per cent and therefore is commercially more important than the ginger-grass oil which contains 19-48 per cent geraniol.

Palmarosa is a native in India, growing wild in the forests. The wild stock has been the major source of oil but it is not of good quality. Moreover, the production of Palmarosa grass in the forest areas has decreased considerably. Therefore the cultivation of this grass for commercial purposes is absolutely necessary. This measure will help to meet the increasing demand for the oil and can be grown in areas where other crops will not give profitable results. Once planted, it can go for 8 years or even more, and the grass can be harvested three times annually from each bush.

The oil is used as a base for perfumes and in cosmetics, particularly in soaps. It is used for flavouring tobacco products and as an ingredient for mosquito repellents. Its use in medicine is noteworthy. India is a major producer and exporter of Palmarosa oil and occupies the third place in the world so far as the export of essential oil is concerned. The country, till recently, had a monopoly over the Palmarosa-oil trade, next to Lemon-grass oil and Sandalwood oil. There is no grazing problem except in the tender stage. The Regional Research Laboratory Bhubaneswar has done extensive pioneering work in the cultivation of different essential oil-yielding grasses and plants, and the distillation of oil. The analysis shows that the net profit per hectare per year comes to around Rs. 8000.

Lemon grass oil (*Genus Cymbopogon*)

Lemon grass is one of the important aromatic grasses yielding essential oil. There are two species namely *Cymbopogon flexuosus* (Steud.) Wats and *Cymbopogon citratus* (DC.) Stapf. The C. flexuosus is known as east Indian Lemon grass which comprises two types of plants. The stem of those plants, having pale purplish colour, is the true Lemon grass and its oil is superior to that of the white stemmed type. The name Lemon grass is given to these types of grasses because the oil, which is distilled from these types, contains citral of 70-80 per cent and above. Lemon grass oil is obtained by steam or hydro-distillation. The oil has a strong lemon odour and is used extensively in cosmetics, particularly in soaps and detergents. A number of other perfumery materials are made from citral.

Lemon grass oil is the second most important EFP item of export, next to Sandalwood oil. There is a great demand for ionones which is obtained through conversion of citral. The ionone is synthesised to vitamin A. This article of perfumery has a great export demand. There is a plant of drugs and pharmaceuticals established at Allepey, Kerala which produces vitamin A from Lemon grass oil.

The Regional Research Laboratory Bhubaneswar has done extensive studies on Lemon grass. The analysis shows that a person can get more than Rs. 4000 as net profit per hectare of land under Lemon grass cultivation.

Tasar-silkworm rearing

The tribal areas, particularly those which are located in and around the forests, are endowed with great potentialities and scope for sericulture. Of all the branches of sericulture the tribal areas of Orissa have ample scope for Tasar-silkworm rearing. It is a large-scale cottage industry and provides employment for plantation of host plants, rearing of silkworms, and reeling and spinning of silk fibres. Tropical tasar-silk is spun by Tasar silkworm (*Antheraea mylitta*), which is the wild non-mulberry silkworm reared outdoor mostly on forest trees and plants like *Shorea robusta* Gaertn.f. (Sal) and *Terminalia tomentosa* (Roxb. ex DC.) Wt. & Arn. (Asan).

Being a wild culture with outdoor rearing, it favours the tribal population to pursue the activities in their forest abode. A large number of tribal families are associated with it either in collection of nature-grown cocoons, or rearing the same on the forest flora. The practice of rearing on the forest trees, which is in vogue in Mayurbhanj district, is called Adapahi. By this practice, the tribal families are allowed to rear silkworms on Sal and Asan trees. Rearing silk worm in *Terminalia arjuna* (Roxb. ex. DC.) Wt. & Arn. has been done inland near village sites which are not suitable for paddy cultivation. These plantations were done in recent years under the Inter-State Tasar Project funded by the Swiss Development Co-operation Mission, which has been wound up since long. Lacking maintenance and care the plantations have suffered a great deal, and the yield has diminished considerably causing a loss and distress to the rearers. The economics of Tasar-silkworm rearing in 1 hectare of land, having about 3500 Asan/ Arjun trees, is given below.

Economics of raising host plants and rearing Tasar silkworms in 1 hectare of land having 3500 Asan/Arjun trees

1st year to 5 the year

(a)	Raising seedlings (Rs. 6,000 per ha)	Rs. 6,000.00
(b)	Plantation of seedlings (Rs. 50,000 per ha)	Rs. 50,000.00
(c)	Maintenance cost (Rs. 2 per plant × 3500 plants per ha × 4 years)	Rs. 28,000.00
(d)	One dug well	Rs. 50,000.00
(e)	One pump set	Rs. 20,000.00
(f)	Diesel and maintenance (Rs. 2500 × 5 years)	Rs. 12,500.00
(g)	Watching the crop cycle - three harvesting period (200 mandays × Rs. 25 per day)	Rs. 5,000.00
(h)	Seeds (Rs. 1,000 per ha)	Rs. 1,000.00
(i)	Implements and accessories	Rs. 20,000.00
(j)	Overhead and misc. expenses for 5 years	Rs. 7,500.00
	Total	**Rs. 2,00,000.00**

The plants will start yielding from the sixth year in the case of Asan trees and from the fourth year in the case of Arjun trees. It has been the experience of the people that although the bushy Arjun plants grow fast and are the best food plants, they do not thrive well in the area. It may be so because irrigation facilities are lacking. With the provisions for irrigation, the situation may improve. Neverthless, both the types of plants should be planted and maintained. In fact, the hectarage under a Arjun plantation is more than under a Asan plantation and both types of plantations are being used for rearing. Therefore, the economic hardships, caused by the longer gestation period of Asan plants, is compensated by the economic benefit derived from the shorter gestation period of Arjun plants.

Tasar silkworm has so far been known to be bivoltine with two crops per year, first one — the seed crop and second the commercial crop. With the introduction of the trivoltine Sukinda breed as also the isolation of some trivoltine strains of Daba, it is now possible to rear a third crop or second commercial crop, capable of generating some additional income. The third-crop's rearing time being October-November/December or cooler months with drier climate, this rearing has had more universal success with lesser attack of pests and predators and less disease mortality.

The plants will start yielding from the sixth year for a period of 50 years provided the plants are properly taken care of by the application of fertiliser, irrigation and pesticides. The yield per hectare of plantation yearly is as follows.

1st harvesting (July-August) 40 days	2nd harvesting (Sept-Oct) 50 days	3rd harvesting (Oct/Nov-Dec) 70 days
4 Kahan × Rs. 400 = Rs. 1600	4 Kahans × Rs. 500 = Rs. 2000	8 Kahans × Rs. 750 = Rs. 6000

Thus a farmer will get Rs. 9600, being the sale proceeds of 16 Kahans (1 Kahan = 1600 cocoons) of cocoons. In addition, he will get fuel worth Rs. 400, which brings a total annual income of Rs. 10,000 from 3500 Asan trees per hectare of land.

Essential forest produce and industries

The volume and variety of EFP available in Orissa forests are still unknown. Whatever is officially procured with respect to all items of

EFP is only a small fraction of the potential yield of such products. The items are not collected to the fullest extent. A major portion finds no market outlet and a large bulk of many items pass out of formal channels by pilferage and clandestine methods. No data are available regarding the quantum of availability of different types of EFP areawise, because no survey or inventory has yet been done. Lacking such data, no entrepreneur nor any funding or lending agency would be willing to invest in any EFP-linked industry. Moreover, non-availability of data regarding what items are available in which area, and in what quantities renders it difficult to take a decision in the matter of location of industries.

There is a list of industries in the report of Decade of Forestry in Orissa, 1981, which are in existence at various places in Orissa. Those units, which use EFP items are given in Table 5.2.

Table 5.2: Industries using MFP (EFP)

	Name of the Unit	*Item of MFP(EFP) used*
1.	Ambica Vegetable and Chemical Private Ltd., Dunguripalli, Bolangir.	Sal seeds
2.	Orissa Vegetable Oil Companies Ltd. Kesinga, Kalahandi.	Sal seeds
3.	Utkal Oil Ltd., Ambaguda, Koraput	Sal seeds and other MFP(EFP)
4.	Orissa Oil Industries, Rairangpur, Mayurbhanj	MFP(EFP)
5.	Utkal Forest Products, Ainthapalli, Sambalpur	MFP(EFP), Harida, Chakunda and Gila.
Forest-based industries (Small-scale)		
1.	Bhattar Solvent, Extraction Udyog (P) Ltd. Arada bazar, Balasore	Sal seed
2.	Frespo Food Products, Jagatpur, Cuttack	Tamarind juice
3.	Jagannath Food Products, Jagatpur, Cuttack	Tamarind powder
4.	Samaleswari Cashew Products Jagatpur, Cuttack	Cashew nuts processing
5.	Agarwala Chemicals, Jagatpur Industrial Estate Jagatpur, Cuttack	Hydrated Castor Oil
6.	Chatrapur Cashew-nuts processing and marketing, Ganjam	Cashew nuts processing

7.	Jain Oil mill, Khariar, Kalahandi	Processing of non-edible oil (Neem/Banatulsi, etc.)
8.	Sitaram Industries, Teliguda, Jeypur, Koraput	Cashew nuts Processing
9.	Munnilal Ramdayal Distilleries, Jugapur, Mayurbhanj	Rectified Spirit
10.	Arati Gum Products (P) Ltd., Mancheswar Industrial Estate. Bhubaneswar	Bubble Gum (Used edible gum)
11.	Falcon Agro Industries (P) Ltd., Tankapani Road, Bhubaneswar	Cashew nuts processing
12.	Kamal wood products, Deogarh, Sambalpur	Katha & Cutch from Khair trees
13.	Hill-broom processing unit, Kashipur, Koraput	Hill-broom
14.	Hill-broom Unit, Naria Road, Kalahandi	Hill-broom
15.	Hill-broom Unit, Thumba, Ganjam	Hill-broom

In addition to those mentioned above, there are many more units such as leaf-plate/cup making, oil extracting, Ayurvedic Pharmacy, Cane-furniture making, Sabai grass rope making, Tasar silkworm rearing, Tasar silk weaving and many other units producing objects of daily use, out of gum, resin, oil seeds, herbs, flowers, fruits, barks, roots and other parts of plants. These few industrial units do not seem to have any impact on the tribal communities for their economic improvements. The reasons are that these are largely private enterprises and self-centred having very little concern for tribal development.

The tribal people continue to serve as wage earners and not as owners of any EFP units. No co-operative societies worthy of note are in existence in any tribal area. Lacking these organisation the tribal communities are unable to imporve their economic condition. It is necessary that primary societies should be established among tribal communities so that they get ample benefit from this source.

6

Policy considerations on EFP Prospects for Development

THE Resolution of National Forest Policy, 1988 lays down the following considerations which govern the establishment of forest-based industries and supply of raw materials to them. The Annexure XVII gives possibility of utilisation of Essential Forest Produce for industrial development having prospects of value addition.

As far as possible, the forest-based industry should raise the raw materials needed for meeting its own requirements, preferably by establishing direct relationship between the industry and the individuals who can grow the raw materials by supporting them with inputs including technical advice and credit needed for harvesting and transport services.

No forest-based industry, except that at the village level, should be permitted to function unless the availability of raw materials is assured. In any case, the fuel, fodder and timber requirements of the local population should not be sacrificed for this purpose.

The forest-based industries must not only provide employment to the local people on a priority basis, but also involve them fully in raising trees and processing the raw materials. Natural forests serve as a gene pool resource and help to maintain their survival. Such forests will not, therefore, be made available to industries for undertaking any plantation by cutting down the existing forests.

The farmers, particularly small and marginal farmers would be encouraged to grow wood species required for industries on marginal/degraded lands available with them. These species may also be grown along with fuel and fodder species on community lands or lands used for pasture purposes.

The study shows that many items of EFP can be processed at the tribal level. The chapter dealing with product profile includes only a few examples of items that can be taken up at the level of primary societies organised by the tribal communities at the grass-root level. Steps are suggested to be taken up for value addition like broom-stick making, and leaf-plate making as well as those requiring complicated processing as in the case of gum, lac, and many such other items of EFP. The Annexure XVII which is referred to above gives a more detailed picture of the utilisation of forest products with particular reference to processing of the products at the grass-root level.

Institutional finance

The tribal communities live in a world of reciprocity and co-operation. The habitat in which they live, the works they do for their livelihood, the festivities which they observe round the year and the crisis which they face in life, compel the tribal people to live a cooperative life. They may not be able to appreciate the official conventional cooperative structure, but a cooperative lifestyle and a cooperative spirit permeate their entire social and cultural fabric.

The tribal society is devoid of any caste and class system; it is undifferentiated, particularly in terms of social hierarchy and economic status. It is largely an egalitarian society having equal levels in wealth and wants across the family, lineage and the tribe at large. Living in an environ subjected to vagaries of weather and in a world peopled by supernatural beings, the everyday life of tribal people is confronted with struggle for existence. They solve their problems, meet their wants, mitigate their sorrows and enjoy their life by controlling the malevolent supernatural forces by harnessing the natural endowments through rituals and hard toil. In a situation of this type, one can imagine how irrelevant the conventional cooperative structure full of complicated and cumbersome rules and regulations is in the life of the tribal communities.

With the opening up of pristine tribal world through the spread of recently developed communication networks and vehicular traffic and as a result of frequent contacts with the outside world, a process of acculturation has taken place and with it, the process of depletion of the natural resources has also occurred. The traditional tribal life is no longer to be seen, and the same sylvan surrounding, the abode of tree, which is caused by death tapping, should be avoided.

Lac cultivation requires technical training. The tribals should be trained in cultivation of lac and collection of both stick lac and brood lac.

Many EFP-yielding plants need replenishment. For example, *Rauvolfia serpentina* (L.) Benth. ex Kurz (Patalagaruda) Zedory and arrowroot are some of such plants. There should be a replenishment plan for such plants as otherwise in a period of a few years these plants will vanish altogether. (See Fernandes, *et.al.,* 1988)

There should be a procurement planning prepared on scientific basis giving remunerative prices as well as sale prices.

Some items are collected individually and some collectively. *Mahua* flower is an example of individual collection whereas hill-brooms need combined effort of an entire tribal village. Such items requiring group labour can be taken up for collection on a co-operative basis. The LAMPS can be made effective by giving responsibility to organise such collective group organisations. (See Saxena, 1995).

Lac collection is dwindling in the State, the reason being lack of interest by procurement agencies. Local traders have been very active in lac trading. A support price including the supply of good quality lac seeds are likely to help in the revival of languishing lac trade. (See report on Marketing of MFP, Government of India, Ministry of Home Affairs, Administrative Staff College, Hyderabad).

There may be monopoly procurement and nationalisation of a few EFP on the grounds of greater profitability by such methods of procurement. But such policy considerations should not overlook the role which the tribals can play in marketing such produce through their own organisation. There is no doubt that the national interest may be taken into account in the matters of monopoly procurement. But at the same time the tribal co-operative societies should be given a chance to associate themselves with Government agencies in procurement and marketing.

The private traders make regular visits to tribal villages and collect EFP from the tribals right at their doorstep. The Government agencies, on the other hand, expect the collectors to come to their depots or procurement centres for disposal of their produce which does not happen on many occasions, thereby the quantity collected by these the archaic tribal life has been vanishing. New ideas have penetrated into the tribal mind and the economy has been monetised, commercial and competitive activities have percolated into these erstwhile tribal areas. The ideas about profit and loss, which are now uppermost in the mind of the tribal people, have replaced the myths and folk tales describing the life of the yester years.

Concerned about the improvement in tribal life, and particularly development of their economic life, numberless measures have been taken since British times. The most noteworthy development strategies, have been evolved and are being evolved from time to time in the Five Year Plans. In order to protect the tribal people from the clutches of moneylenders, merchants and traders, many co-operative credit societies, fair price shops and many other categories of co-operative and commercial institutions have been set up in the tribal areas to provide institutional finance and credit to the tribal people and supply their daily necessities and buy from them their SAP and EFP at a reasonable price.

There are now 25 Commercial Banks and 9 Regional Rural Banks operating in the state of Orissa. The branches of commercial banks, which are now 2131 in number, are in operation in different districts. Compared with the coastal districts, the hilly areas which are dominated by the tribal communities are covered with less such banks and Cooperative Societies in terms of population and area coverage. The area coverage, which stands on an average for the state at 73 sq km per bank, exceeds many times in the case of tribal districts.

Considering uncertainty in achieving a higher density of banks, alternatives have been tried to meet the requirements of credit. One of such possibilities is organisation of self-help groups in tribal villages. In some places trials have been given financial assistance through bank for the self-help groups to operate, and results in this regard have been very much satisfactory. These self-help groups promise to place the banking facilities closer to the people, help people to get maximum benefit with minimum costs in terms of meeting the credit requirements both for production and consumption needs of the members. Similarly, community credit facilitators are thought of as another alternative who would meet the credit requirements of beneficiaries in a block under their close supervision.

As regards the co-operative credit system, there are 17 co-operative credit banks affiliated to the Orissa State Co-operative Bank and there are 2817 Primary Agriculture Co-operative Societies including 223 LAMPS operating in the tribal areas. As is well known, the functioning of LAMPS is not satisfactory. Some of the reasons for the failure of LAMPS as reviewed by the Working Group on Development and Welfare of STs during Eighth Five-Year Plan Period (1990-5) are :

- Non-viable areas of operation and activities of LAMPS,
- Frequent supersession of elected bodies and appointment of official management,
- Continuance of co-operatives like PACs and FSSs along with LAMPS,
- Non-repayment of loans,
- Lack of own sources with LAMPS.

Moreover there are some other reasons. They are :

- Heavy royalties on Essential Forest Produce,
- Inadequate margins on consumer business in essential commodities,
- Heavy cost of transport and handling of TSP areas,
- Procedural requirements required to be fulfilled by loanees, leading to delay in disbursement of loans, and
- Lack of refinancing of certain activities like trade in EFP, and in essential commodities and consumption loans as NABARD does not presently have lines of credit for such purposes.

The analysis of these points in terms of political issues shows :

- Area-wise and activity wise the LAMPS are not viable,
- The politicians for their vested interest and economic gain are in the habit of changing elected bodies and appointing offical staff of their own choice,
- Allowing the PACs and FSSs to continue side by side the LAMPS hampers progress and development of the LAMPS,

- Conflicting ideas and struggle for power among political parties stand in the way of gaining economic strength among the LAMPS and in such a situation the LAMPS become weak and incapacitated in performing their responsibility in a proper way,
- The loanees do not pay back the loans they received in time and therefore the LAMPS become defunct, and
- The LAMPS meet heavy costs of transport of goods and services with which they deal and have to pay heavy royalty or Essential forest produce with which they carry on their business.

The study undertaken by the NABARD has also pointed out the failure of LAMPS for some different reasons. Some of the findings are :

- Very large areas covered under LAMPS making them ineffective to meet the demands of the tribals,
- Heavy transportation expenditure involved in carrying goods from purchasing centres to branch offices/head offices,
- The financial condition of the LAMPS is very weak. They are unable to purchase required quantities of fertilisers in time to meet the demands of the farmers and financial constraints of making them ineffective in running day to day business,
- Most of the LAMPS are running at loss to such an extent that they are not in a position to meet even the salaries of the staff members,
- The transport charges paid by the Orissa State Co-operative Marketing Federation Ltd. is not sufficient to meet the expenditure incurred by LAMPS towards transportation of fertilizers, LAMPS incur loss as shortages of fertilisers occur in retain sales,
- Moreover, the LAMPS are supposed to act as agents of the Tribal Development Co-operative Corporation of Orissa Ltd. (TDCC) in procuring Essential Forest Produce. But, the TDCC does not appoint all the LAMPS as its agents and even those appointed are not given agency to procure all types of Essential Forest Produce.

Added to these reasons, co-ordination and healthy organic linkage between different tiers of co-operative marketing structure (both horizontal and vertical) viz., LAMPS, TDCC, RMCS, OSCMF, TRIFED, NAFED are lacking, hence difficulties in proper functioning of the LAMPS. Moreover, the share capital and managerial assistance is not adequate and hence most of the LAMPS borrow funds from the financing agencies to undertake consumer business. The financing agencies charge very high rates of interest to the extent of 13% and above. Due to less margin and high managerial expenses, the LAMPS are not able to repay the bank loans and their viability is affected. Officially the staff members of LAMPS do not belong to a single cadre. They are brought on deputation basis from Central Co-operative Banks, Co-operation Department, TDCC etc. Even for the staff of the same ranks, different scales of pay are paid. The salary variations among the staff affect the work efficiency to a large extent.

Taking into account the information based on NABARD's study the working groups made the following recommendation as matters of policy.

- The monitoring of flow of credit through banks by the Reserve Bank should be improved so as to give the flow of credit to SCs and STs separately and also to make it possible to assess the quantum of credit flowing to STs in a given period of time.
- The licences issued for opening branch offices of banks should be implemented.
- Considering the low density of population in many of the TSP areas, a survey should be carried out of the actual number of branches required as per norms already prescribed. This survey could be taken up by the Ministry of Finance through NABARD and the Agricultural Finance Corporation.

Moreover some other recommendations are :

- The District Consultative Committees and Standing Committees should monitor the flow of credit to STs/TSP areas, and the Reserve Bank of India may issue instructions in this regard.

- No lending bank should ask for collateral security or prescribe a sub-optional repayment schedule in violation of the guidelines on the subject.
- Necessary training is imparted to bank personnel posted in TSP areas in preparing bankable schemes appropriate to local conditions.
- The credit under government-sponsored schemes as well as short-term agricultural credit should include a consumption finance component to enable the ST families to tide over the lean period. Consumption loan facility should also be available for religious/social needs.
- In respect of production credit and consumption credit, separate limits should be sanctioned by the lending bank/ cooperative and the loanee should be permitted to operate a running account, drawing and repaying at his convenience.

Added to these some other points are that :

- NABARD does not have a line of credit now for refinancing the consumption finance requirements of LAMPS or other Co-operatives. Such a line of credit should be established.
- Supersession of elected bodies in LAMPS or other cooperatives in tribal areas should not be resorted to, and the cooperatives should be allowed to recruit their own employees as far as possible.
- The district level federation of LAMPS may be established, at least in the full ITDP districts, to strengthen the bargaining position of the LAMPS.
- The system of monitoring of flow of credit to STs through cooperatives should be improved substantially by NABARD.
- All ST families having need for credit should be enrolled as members of co-operatives through provision of grants towards share capital.
- In the TSP areas, all posts of chairman/president of co-operatives and 2/3rd of the elected bodies should be reserved in favour of STs.

- Necessary legal provision should be made for reservation of posts other than MD/Paid Secretary of the co-operatives in TSP areas in favour of the local STs in proportion to the percentage of ST population to total population of the ITDP in which the co-operative is situated.
- By way of suggestions the working group pointed out that in order to make the co-operatives viable, the margin should be increased on the controlled commodities, and refinancing should be made available at concessional rates of interest for marketing of EFP, agricultural produce, business in controlled commodities and agricultural inputs. A separate line of credit would have to be made available to NABARD for this purpose.
- Since a large number of cooperatives in TSP areas have become defunct on account of non-repayment of loans, the norms for refinancing of cooperatives by NABARD may be relaxed.
- Full and compulsory credit insurance may be provided to guard the cooperatives against crop failures.
- The employees of cooperatives should be trained at Government cost at point of entry and during service.
- In the case of cooperatives dealing in Essential Forest Produce and Surplus Agricultural Produce, repayment of loans and interest thereon should be accepted in kind against a running account.

7

Policy Assessment and Concluding Remarks

KEEPING the objectives of the study the EFP in Orissa in view, observations were made at the field level, interviews were conducted with officials and non-officials who were connected with EFP in the State, records and published materials were referred to for collection of data on the subject-matter of the study. Primarily the forest policies and policy considerations having bearing on EFP in Orissa were covered in the study and prospects for development as derived from the impact of the forest policies (see chapter 2) and some recommendations which emerged from the study are dealt with in this section.

Data collection and record keeping

It was the observation that although data on all aspects of the subject-matter of the study are available with different agencies, the records and sources of information are not properly and systematically maintained, and therefore, data collection is bound to take time. It is necessary that a system has to be evolved for record-keeping and compilation of data in a systematic manner.

As there is no exploratory and inventory survey done in the State, a total picture of EFP available in the forests of Orissa is not known officially. From time to time, new items are found out and collection is taken up. Sometimes some external agencies are in search of certain items which are known to them but unknown in the official circles, for example the barks of some trees are used to make *agarbati* (incense sticks). These external agencies get these items procured, unknown to the official circle by employing suitable collectors who are able to identify these items.

In order to facilitate the proper collection of EFP it is necessary to have a map showing the distribution of different items of EFP in different areas of the state. This map will also indicate seasons in which different items of EFP are available and the communities which are collecting such items. A resource survey with inventory of EFP helps to organise procurement activities well in advance.

All items of EFP should be listed for collection, the utility of all these items should be found out and adult education should form an essential part of the procurement planning for an optimum collection of remunerative items.

Different items of EFP are available in different seasons in the natural setting and this aspect of seasonality is most important in the collection of medicinal plants. The medicinal value of many plants or parts of those plants is lost unless these items are not collected in proper time. Therefore, there should be a seasonal calendar showing a list of medicinal plants and the seasons of collection.

The next items are related to collection, procurement and processing strategies. Some of the observations and facts are given below in this regard.

Collection and procurement strategy

Current levels of collection are far below the availability of EFP because of the following factors:

- Inaccessibility of the areas and hilly terrain for the tribal communities in certain seasons, particularly in the case of tribal women who do the collection more than the men.
- Pilferage and clandestine collection of EFP by unauthorised persons.
- Short duration of availability of certain EFP such as *Sal* seeds, *Kendu* leaves, etc.
- Perishability of certain items of EFP, particularly ripe fruits such as *Kendu* (fruit of *Diospyros melanoxylon* Roxb.), *Jamun* (fruit of *Syzygium cumini* (L.) Skeels.) *Khajuri koli* (fruit of *Phoenix sylvestris* (L.) Roxb.

Tribals should be trained in proper methods of collection of EFP. It is particularly necessary for the extraction of gum. An injury to the

Sl. No.	Division	District	R.F.	D.P.F.	U.D.P.F.	U.C.F.	Total
8.	Baliguda	Phulbani	1029.52	1173.94	564.98	0.59	2769.03
9.	Boudh	Phulbani	1015.14	32.68	57.94	1.02	1116.78
10.	Gh. North	Ganjam	909.69	8.14	—	0.53	918.36
11.	Gh. South	Ganjam	530.23	517.62	697.74	0.30	1745.89
12.	Nayagarh	Puri	1098.37	385.28	310.45	0.08	1794.18
13.	Phulbani	Phulbani	980.54	658.51	124.84	1.40	1765.29
14.	Puri	Puri	433.30	191.01	692.88	1.37	1318.56
	do	Cuttack	—	—	5.63	0.01	5.64
15.	Parlakhemundi	Ganjam	446.12	2053.18	1869.17	0.16	4368.63
16.	Bolangir	Bolangir	1203.37	30.50	—	0.16	1234.03
17.	Jeypore	Koraput	480.36	855.83	754.67	0.53	2091.39
18.	Kalahandi	Kalahandi	1448.83	864.33	1537.22	0.54	3850.22
19.	Khariar	Kalahandi	—	1663.53	139.40	0.44	1803.37
	do	Sambalpur	22.40	472.30	125.47	0.07	620.24
20.	Nawarangpur	Koraput	584.08	1152.36	19.11	0.12	1755.67
21.	Rayagada	Koraput	1079.72	2273.47	1666.36	0.18	5019.73
22.	Bamara	Sambalpur	722.91	380.50	426.97	0.95	1531.38
23.	Bonai	Sundergarh	1162.64	432.44	237.54	0.71	1833.33
24.	Deogarh	Sambalpur	612.97	185.25	169.64	0.02	967.88
	do	Dhenkanal	289.69	49.52	49.52	0.01	388.74
25.	Rairakhol	Sambalpur	1013.39	126.66	484.30	0.51	1624.86
	do	Bolangir	55.48	—	90.00	—	145.48

agencies does not reach the target fixed by them. But the organisations of co-operative societies among the tribal communities will discourage such malpractices.

Mobile procurement vans and setting up procurement centres in inaccessible areas may increase the volume of EFP collected. This arrangement can be assigned to the co-operative society which will result in greater amount of collection.

Nationalisation and monopoly procurement which means State organised trading mechanisms have as much merits or demerits as privatisation and the free market. Even in the legal monopoly system, the gatherers are paid less than the fixed rate and agents go scot free. Sometimes the agents insist on barter payment in the form of grain. Just before the collection season in place of previous arrangements of procurement by leaseholders, private parties/ industries are appointed. As a result, the primary gatherers suffer, because the newly appointed agencies, who are not ready to deploy men and money for the procurement. They also do not turn up in the field in time, and therefore, the gatherers are confronted with a distress sale situation and dispose of their produce to the middlemen and local traders at a throwaway price. The monopoly system may give guarantee for payment of royalty regularly. But it is found that the procuring agencies resort to deferred payment while buying the products from the gatherers and also fail to pay the revenue to the Government regularly. All these drawbacks can very well be tackled if the co-operative societies are established and all facilities and encouragements given to the societies for functioning satisfactorily.The policy frame should appreciate the need to reestablish the rights of the tribals for the collection of EFP items. In fact, the National Forest Policy, 1988 has emphasised this matter and has been most favourable in safeguarding the rights of the tribals over EFP.

The private traders make payments of royalty in advance lest they may not get the lease. But one disadvantage is that the tribal gatherers accept whatever price the private traders pay. However, the criteria for decisions on procurement policy should be :

(a) Importance to the tribal economy and

(b) Potential for value addition.

Procurement planning should cover among other items such prerequisites as storage facilities with scientific storage techniques, link roads, approach roads, transport facilities, system of marketing, processing and organisational set-up at various levels. Systematic procurement and proper utilisation of EFP for economic development require collection centres at strategic places, free flow of credit, location of banks, co-operative societies at proper places, industries and processing units and such other services.

Processing is essential to add value to EFP to benefit tribal people who are primary gatherers of such items. Most of the products can be processed at the preliminary stage right in the houses of the tribal people. Very simple and less expensive decorticators are available which the tribals can have for deseeding and decorticating tamarind seeds, *sal* seeds, *mahua* seeds, etc., and sell kernels at a higher price. Binding broom grasses to make broomsticks is an easy task. Cultivation of aromatic grasses and distillation of essential oils which require group efforts can be taken up in the tribal villages. A herbarium can be raised at the family level or collectively among a group of families. The making of leaf cups/plates out of *Sal* leaves can be taken up on a co-operative basis.

Sabai-grass rope-making is an important value addition, and rope-making can be taken up either at the individual family level or collectively. Tasar-silkworm rearing and lac cultivation are some of the traditional occupations of the tribal communities who have developed great skill in producing tasar cocoons and stick lac. These activities are ideal for value addition of higher order which fetches great benefit for the tribals. Scientific training in tapping gum-yielding plants, quality specifications and grading techniques will not only benefit the tribals through value addition, but also save the gum yielding species from extinction which result by death tapping and excessive blazing of the plants to exude gum without rest. (N. Patnaik's own experience gained in Mayurbhanj Pilot Project, sponsored by ILO).

The Regional Research Laboratory (RRL) is equipped with technical know-how regarding the extraction of essential oils from aromatic grasses, plants and spices. Similarly technical know how about the extraction of medicinal alkaloids and acids from some items of EFP is available there. This technical knowledge can be utilised by the co-operative agencies and Governmental undertakings for processing specific items of EFP.

Bottling and canning of many edible plums and fruits do not need any complicated processing. Pickles and jams can be prepared with ease out of *Emblica officinalis* Gaertn. (Amla), *Tamarindus indica* L. (Tentuli), *Mangifera indica* Linn. (Amba), *Ficus glomerata* Roxb. (Dimiri), *Syzygium cumini* (L.) Skeels (Jamun), *Diospyros melanoxylon* Roxb. (Kendu), *Aegle marmelos* (Linn.) Corr. (Bela), *Phoenix sylvestris* (L.) Roxb. (Khajuri), etc. Honey, arrowroot and *Amorphophallus campanulatus* Bl. (Olua) can be processed at the village level and molasses can be prepared from *Borassus flabellifer* Linn. (Tal). There are many EFP items of great food value which can be processed under training-cum-production schemes.

It may be noted that most of these schemes are seasonal, their availability is limited to short periods. It is, therefore, necessary to have an efficient procurement system with subsidy and price support incentives, fast-moving transport services, and scientific preservation and storage facilities so that these perishable products can be processed when they are in good condition.

Both collection and procurement of EFP are basically economic activities of the State. This activity is also linked with the livelihood of the tribal people. But these points have implications for tribal development. The policy will be such that the activities should not rob the tribals of the items which they are presently utilising for themselves. For instance, items like *sal* seeds and *mahua* seeds should be utilised first to meet the needs of tribal people and thereafter further processing and trading may be taken up.

Greater economic gain to the tribal communities can be achieved by establishing tribal co-operative societies which will be entrusted not only with collection and procurement but also with processing of the EFP. Until such time when decisions are taken at the policy level some via media arrangements may be made for greater involvement of the tribal communities in different aspects of EFP.

Setting up of tanneries, production of edible oil and non-edible fats and production of soaps should be taken up in the tribal areas so that the tribal people can afford to buy the finished products for their own consumption in the place of costly and fancy items which they cannot afford to buy. Even in such matters the tribals can be involved through their co-operative organisation with necessary support from the Government.

The problem of value addition to benefit the tribal people is crucial in the study of EFP. Depending on the simplicity or complexity of processing, it may be done broadly at three levels — tribal household or primary co-operative level, group of co-operatives level, and LAMPS or higher level. The benefit of value addition of the first stage should accrue directly to the tribal households. Value additions at the other two stages are to be passed on to the tribals as a profit or as a grant to their co-operative organisations to be shared by individual members.

As regards the processing at the higher stages, the benefits may flow to the tribals through grants placed within the development set-up of the Tribal Sub-plan. At every stage, there are dangers of benefits being siphoned off by private parties. Therefore, all care have to be taken and all steps should be so arranged and their marketing so planned that the increased value falls in the hands of the tribal people. The Government of India has specific recommendations on EFP through various working groups set-up in different Five Year Plans.

Pricing, marketing and organisation are some of the most important aspects of EFP. These matters are dealt with below briefly.

Inter-State coordination in respect of price fixation is essential; otherwise it will encourage smuggling. The main purpose of the Government is to ensure a fair return to the tribals for the items of EFP they collect. To achieve what is needed is that the rights of the tribals to collect should be codified and made public, and greater involvement of the tribals through collectors' primary co-operative societies and training in scientific collection of materials through non-formal education should be encouraged to help the tribal people to derive greater benefit from the collection of EFP.

The main guiding principle should be a fair return to the tribals. Sound norms of costing of performances of different activities from the stage of collection to the stage of disposal will be the guiding principle for determining the fair return to the tribals. A specialised body may be set-up to identify the components of cost for different activities connected with various items of produce, and determine both procurement and sale price of each item of EFP. Royalties should be fixed after examining the nature of contributions and cost involved for various services associated with EFP. It will be considered to make the royalty fully or partially subsidised so that the benefit flows to the tribal communities. A kind of guarantee of a minimum price may be fixed for different items of EFP as a measure to protect the tribal people from the vagaries of market.

Any loss incurred by the procurement agencies on account of this protection, the loss should be reimbursed to them by the Government. The Government should not compensate costs incurred by the Corporation in such situations as operational inefficiencies. By doing so, incompetence of the Corporation grows rather than is corrected. The products turned out by the processing units utilising the EFP should be exempted from excise and other taxes.

The State has some definite role to play in this matter of marketing the EFP. It would help the societies to establish linkages with TDCC, OFDC, ORMAS, LAMPS which will buy the EFP from the co-operative societies for marketing purposes. These co-operative societies should be fed with information from time to time regarding the potential of different products for commercial purposes so that they give proper attention to different products at the time of collection and procurement and storage.

The grass-root level organisation should be trained to have in course of time necessary capability of handling administrative, commercial, financial and other problems. To make this arrangement effective in the matter of giving the tribals a maximum return from the produce they collect there should be a two-tier organisation – the primary tier with village voluntary self-help groups and Mahila Mandals as members at the grass-root level and the apex tier a confederacy of primary tiers as members at the inter-village level. The primary tier should have responsibility of collection and procurement, the apex tier having the responsibility of pricing, marketing, processing, storage, etc. The members of the apex tier will have further responsibility of forming a cadre of tribal youths and providing training to the youths to man the positions in LAMPS and establishing linkage with the Forest Department for expert advice and guidance. The success of this organisation depends upon the personnel holding the job assignment connected with the establishment of such an organisation at the grass-root level. This work needs suitable staff who are dedicated to work for tribal welfare and who know how the tribals are linked with forest and who is trusted by the tribals.

The above-mentioned aspects and materials show how complex the study on EFP is. It is multifaceted and multidisciplinary and these are interrelated. Unless a holistic view of this subject-matter is taken the study of one aspect without keeping the interlocking aspects in view the findings appear to be disjointed and incomplete. Therefore it is necessary to undertake a scientific research on different aspects before any attempt is made to present a picture of this complex subject. The last paragraph gives the points which require investigation with the study on this subject.

Research and development

Research and development should form an integral part of the collection, procurement, storage, marketing and processing package of EFP. All information regarding these activities should be disseminated to all agencies concerned with EFP.

There are many agencies active in research in this area. The research activities which are carried out by these agencies are all independent, uncoordinated and unplanned. The Forest Research Institute has done a good deal of work regarding commercial utilisation of EFP. This institute will be an ideal body to co-ordinate all research and development activities concerning potentials of EFP, planning, afforestation for renewing EFP resources and develop plantation of crops for EFP.

The works connected with tanning and tanneries are done by the Central Leather Research Institute, Madras. The institute has the technical know-how regarding utilisation of tannin yielding materials and setting up of tanneries. The KVIC can also help in setting up such industries.

The RRL Bhubaneswar has developed works connected with dyes and extraction of aromatic oils. The Central Food Technological Research Institute, Mysore can undertake all works connected with edible oils, animal feed from the by-products of non-edible oilseeds. The Dairy Research Institute at Karnal can be entrusted with works connected with the development of cocoa butter from mango kernel and *sal* seeds. The RRLs and Central Indian Medicinal Plant Organisations (CIMPO), Lucknow and National Botanical Garden, Lucknow have been working in the area of medicinal plants. Similar research works are also being undertaken by the Central Council for Research in Indian Medicine and Homeopathy (CCRIMH), New Delhi. These organisations can be of great help in carrying out research in herbal drugs.

More than 1500 non-governmental organisations (NGOs) from all over the country have presented a jointly-prepared draft Forest Bill to the Ministry of Environment and Forest as an alternative to the Bill formulated by the Ministry.

The main criticism of the draft official Bill is that it focuses upon conservation alone, overlooking the needs of the people, although the National Forest Policy, 1988 envisages to make the programmes of afforestation as a people's movement. The alternative Forest Bill focuses upon a partnership between the village committee and the Forest Department. The effort would be incorporated in the alternative Forest Bill, concentrated upon turning the forest management into an income-generating programme and poverty eradication activity. The Annexure XV presents the views of the NGOs seeking changes in the official Forest Bill.

As pointed out earlier the Essential Forest Produce is a multi-disciplinary and multifaceted study. A comprehensive guideline is needed before the study is taken up. As an example, a general guideline is given in Annexure XVI. The EFP cannot be viewed in isolation from other aspects of forestry. It is an integral part of the forest wealth and forest resources which are most important renewable resources of a country. While the plans are being prepared for afforestation in respect of timber and firewood, the EFP has till now not received any such attention. In fact, there has not only been no protection to the EFP resources leading to indiscriminate destruction of many EFP-yielding species, but also there lacks plans for renewal of resources connected with the EFP. This means that one would have to move away from timber-oriented forestry to holistic forestry in which the rich diversity of our natural forest is replicated in those developed by human effort. (See chapter 16 of *Tribal settlement and Minor Forest Produce* edited by D. Thakur and D.N. Thakur).

Annexure1*

Forest area of territorial divisions as on 01.04.90 by legal status

Sl. No.	Division	District	R.F.	D.P.F.	U.D.P.F.	U.C.F.	Total
1.	Angul	Dhenkanal	907.88	211.06	816.30	0.76	1936.00
2.	Athgarh	Cuttack	428.00	250.37	336.59	0.45	10.15.41
3.	Athamallk	Dhenkanal	527.36	—	629.76	0.42	1157.54
4.	Baripada	Mayurbhanj	383.92	118.54	75.84	0.53	1127.43
					Lakhraj		
					548.60		
	Baripada	Balasore	202.68	3.68	57.37	0.19	263.92
					Ex-Jamindari		
5.	Dhenkanal	Dhenkanal	1140.97	47.60	582.57	0.04	1860.13
					36.34		
					Lakhraj		
					52.61		
					Devottar		
6.	Karanjia	Mayurbhanj	379.61	186.72	145.69	0.27	738.45
					26.29		
					Lakhraj		
7.	Keonjhar	Keonjhar	1833.02	648.41	12.68	0.24	2494.35
	Keonjhar	Cuttack	—	300.71	18.15	0.01	318.87

26.	Sambalpur	Sambalpur	1012.95	51.06	227.35 29.92 Ex. Malgujari 134.80 Ex. Jamindari	0.28	1456.36
	do	Bolangir	57.50	—	—	—	57.50
27.	Sundergarh	Sundergarh	1489.75	470.88	194.10	1.01	2155.74
28.	WLCO Chandaka	Cuttack	—	61.14	12.97	—	74.11
	do	Puri	78.30	31.00	61.14	0.08	170.52
29.	WLCO Chandbali	Cuttack	69.81	150.28	156.97	0.11	3.77.17
	do	Balasore	—	21.00	8.88	—	29.88
30.	S.T.R. Baripada	Mayurbhanj	303.00	—	—	—	303.00
31.	SFDC Baripada	Mayurbhanj	2328.00	169.07	181.39 Corp. bodies	1.45 Corp. bodies	2679.91
32.	CFWP, Circle, Ctc.	Cuttack	—	—	—	0.01	0.01
33.	CF Affr. Bolangir	Bolangir	—	—	—	0.23	0.23
	do	Koraput	—	—	—	0.18	0.18
34.	CF Dev	Dhenkanal	—	—	—	0.02	0.02
35.	FRC. Angul	Dhenkanal	—	—	—	0.06	0.06

* Adapted from *A Decade of Forestry in Orissa, 1981-2*, Office of PCCF, Orissa, pp.11-12.

Annexure I (a)

Different types of MFP(EFP) (Categorywise listing of MFP(EFP) and other information)

Sl. No.	Name of MFP(EFP)	Botanical term	Area where available in abundance	Season of availability	Average annual collection in qtl.	Average procurement value per qtl. in Rs.	Sale value per qtl. in Rs. (1994-5)
Fibres & Flosses							
1.	Antia Fibre	*N.A.*	N.A.	April to May	44.00	91.00	N.A.
2.	China Climber	*N.A.*	Karanjia, Baripada	April to May	80.00	50.00	100.00
3.	Dhantari lata	*N.A.*	Rairakhol, Bonei, Deogarh, Rayagada, Kalahandi.	N.A	1650.00	N.A.	N.A.
4.	Jungle jute	*N.A.*	N.A	N.A.	N.A.	N.A.	N.A
5.	Kathalai	*N.A.*	Parlakhemundi	N.A.	8.85 qtl. & 82.00 pana	1355.93 1.28 per pp	0.36 P.B.
6.	K. Creeper	*N.A.*	Ghumser South, Nayagarh, Puri	April	18683 qtl 11450 kera	450 per qtl 1.34 p.k.	2.72 p.k.
7.	Siali Fibre	*Bauhinia vahlii* Wt. & Arn.	Raygada, Sundergarh, Nowarangpur, Bhawanipatna	April to May	427.06	860.00	990

8.	Simuli cotton	*Bombax ceiba* (Linn.)	Karanjia, Baripada, Sambalpur	April to May	11.35	2000.00	4000.00
9.	Sisal Fibre	*Agave sisalana* Perr. ex Engelm.	N.A.	April to May	230.00	40.00	N.A.
10.	Swan climber	*N.A.*	Karanjia, Baripada	December	317.58	84.00	86.00
Grasses, Reeds, Cane and Fodders							
1.	Cane	*Calamus rotang* Linn.	Athagarh, Puri,	April to May	211 Kahan	132.65 P.ka	242 P.Ka
2.	Dry grass	N.A.	N.A.	N.A.	N.A	N.A.	N.A.
3.	Hill brooms	*Thysanolaena maxima* (Roxb.) Kuntz	Kalahandi, Jeypore, Nowrangpur, G. South Khariar, Rayagada.	Dec. to April	2568.35	433.23	3.11 P.B.
4.	Kantabandhuri	N.A.	Bonai	N.A.	1874.00	N.A.	N.A.
5.	Matreed	N.A.	N.A.	N.A.	93440.00 Pannel	N.A.	N.A.
6.	Sabai grass	*Eulaliopsis binata* (Retz.) Hubbard	Angul, Karanjia, Baripada	Oct.	5200 pana	5.00 P.P.	6.00 P.P.
7.	Thatch grass	N.A.	Karanjia, Baripada	Nov. to Feb	N.A.	N.A.	N.A.
8.	Thorn broom	N.A.	Bhawanipatna	Dec. to March	1864.18	150.00	208.00
Essential or Volatile Oils							
1.	Benachera	*Vetiveria zizanioides* (L.) Nash	Samabalpur, Sundergarh	N.A.	427.88	289.88	N.A.

Annexure I (a) (Contd....)

Sl. No.	*Name of MFP(EFP)*	*Botanical term*	*Area where available in abundance*	*Season of availability*	*Average annual collection in qtl.*	*Average procurement value per qtl. in Rs.*	*Sale value per qtl. in Rs. (1994-5)*
2.	Harida	*Terminalia chebula* Retz.	Korapur, Kalahandi Mayurbhanj, Sundergarh	Nov. to March	12768.77	165.58	218.00
3.	Kamala Gundi	*Mallotus philippensis* (Lam.) Muell.-Arg.	Bolangir, Rayagada	N.A.	25.56	225.00	N.A.
4.	Marking nut	*Semecarpus anacardium* Linn.	Keonjhar, Baripada, Karanjia	Nov. to March	285.00	50.00	132.60
5.	Sunari bark	*Cassia Fistula* Linn.	Nayagarh, Deogarh	April to May	440.20	126.50	165.00
Gums, Resins and Oleoresins							
1.	Amta gum	N.A.	Keonjhar, Karanjia	Jan. to Feb.	1472.50	816.14	1004.00
2.	Babul gum	*Acacia nilotica* (L.) Willd. ex Delile	Kalahandi	May to June and Oct. to Nov.	N.A.	N.A.	N.A.
3.	Bahada gum	*Terminalia bellirica* (Gaertn.) Roxb.	Sundergarh, Keonjhar, Baripada, Karanjia	January to April	442.65	1030.00	1262.50
4.	Char gum	*Buchanania lanzan* Spr.	Sundergarh	January to April	574.00	100.00	140.00
5.	Dhaura gum	*Anogeissus latifolia* (Roxb. ex DC.) Watl. ex Guill & Perr.	N.A.	January	26.98	674.88	N.A.

6.	Genduli gum	*Sterculia urens* Roxb.	Rayagada, Jeypore	January to April	183.00	1690.00	6000.00
7.	Khaira	*Acacia catechu* (L.f.) Willd.	N.A.	N.A.	N.A.	N.A.	N.A.
8.	Sal resin	*Resins of Shorea robusta* Gaertn. f.	Puri, Jeypore, Nowrangpur, Keonjhar	March to June	7.74	1200.00	N.A.
Drugs and Medicines							
1.	Anla	*Emblica officinalis* Gaertn.	Bolangir, Kalahandi Nawarangpur	Nov. to March	42.31	159.46	237.00
2.	Antudi flower	N.A.	Baripada, Keonjhar	April to May	68.25	100.00	185.00
3.	Bahada	*Terminalia bellirica* (Gaertn.) Roxb.	Bolangir, Kalahandi Nowrangpur	Nov. to March	317.00	62.84	85.00
4.	Bala (Bajramuli)	*Sida cordata* (Burm.f.) Brossum	N.A.	N.A.	N.A.	N.A.	N.A.
5.	Banahaldi	*Curcuma aromatica Salisb.*	Bonai, Khariar, Kalahandi	N.A.	0.87	150.00	N.A.
6.	Beherfruit	N.A.	N.A	N.A.	N.A.	N.A.	N.A
7.	Kalmegh	*Andrographis paniculata* (Burm.f.) Wall. ex Nees	Sundergarh	Sept. to Dec.	228.00	100.00	130.00
8.	Dhatuki flower	*Woodfordia fruticosa* (L.) Kurz	Sundergarh, Kalahandi Rairakhol, Khariar, Nowrangpur	Dec. to March	338.17	40.00	228.00
9.	Eksira fruit	N.A.	Keonjhar	N.A.	N.A.	10	100

Annexure I (a) (Contd....)

Sl. No.	Name of MFP(EFP)	Botanical term	Area where available in abundance	Season of availability	Average annual collection in qtl.	Average procurement value per qtl. in Rs.	Sale value per qtl. in Rs. (1994-5)
10.	Fenfena fruit	*Oroxylum indicum* (Linn.) Benth. Ex Kurz	Sambalpur, Bamra, Baripada, Dhenkanal, Deogarh, Parlakhemundi	March	10.00	224.71	410.00
11.	Gunja (Ranju)	*Abrus precatorius* Linn.	Karanjia, Baripada	N.A.	N.A.	N.A.	N.A.
12.	Nirmala	*Strychnos patatorum* L.f.	Baripada, Boani	N.A.	N.A.	N.A.	N.A.
13.	Kurchi bark	N.A.	N.A	N.A.	N.A.	N.A.	N.A.
14.	Landabaguli	*Ocimum basilicum* Linn.	N.A.	N.A.	N.A.	N.A.	N.A.
15.	Lodhamedha	*Symplocos racemosa* Roxb.	Ghumsur (N&S), Bonai, Boudh, Kalahandi.	N.A.	N.A.	N.A.	N.A
16.	Mankada kendu	*Diospyros malabarica* (Desr.) Kostel	Sundergarh, Patnagarh, Jeypore	June to July	12.50	100.00	200.00
17.	Nageswar flower	*Mesua ferrea* Linn.	Mayurbhanj, Deogarh	Jan. to Feb.	3.12	1500.00	1900.00
18.	Nuxvomica	*Strychnos nux vomica* Linn.	Mayurbhanj, Rayagada	Dec. to Feb.	2130.94	662.95	1557.00

19.	Pita alu	*Dioscorea bulbifera* Linn.	Karanjia Baripada	Oct to Dec.	64.25	46.87	250.00
20.	Putrani	N.A.	Nowrangpur	N.A.	N.A.	N.A.	N.A.
21.	Rohini flower	*Mallotus philippensis* (Lam.) Muell. -Arg.	Baripada	Feb. to April	24.00	83.35	220.00
22.	R.S. root	*Rauvolfia Serpentina* (L.) Benth. ex Kurz	Rayagada	Oct. to Jan.	10.50	1289.40	1650.00
23.	Sidha fruit	*Lagerstroemia parviflora* Roxb.	N.A	N.A.	N.A.	N.A.	N.A.
24.	Rasna	*Vanda tessellata* (Roxb.) Hook. ex G. Don	Kalahandi	N.A.	N.A.	N.A.	N.A.
Oil Seeds							
1.	Ananta seeds	*Hemidesmus indicus* (Linn.) R.Br.	Phulbani, Jeypore, Nowrangpur	N.A.	3.63	400.00	N.A.
2.	Babul seeds	*Acacla nilotica* (L.) Willd. ex Delile	Kalahandi, Bolangir, Khariar	N.A.	3340.50	61.86	N.A.
3.	Baghanakhi seeds	*Martynia diandra* Glox.	Bolangir, Phulbani Kalahandi, Bonai	N.A.	6.16	75.88	N.A.
4.	Baidanka Seeds	*Mucuna monosperma* DC. ex Wight	Jeypore, Kalahandi, Bolangir	N.A.	12.17	200.00	N.A.
5.	Bananila Seeds	*Tephrosia purpurea* (L.) Pers.	Jeypore, Nowarangpur, Kalahandi	N.A.	480.46	240.00	N.A.
6.	Banatulsi seeds	*Ocimum gratissimum* Linn.	Sambalpur, Jeypore, Sundergarh, Bolangir, Kalahandi	N.A	7427.70	235.00	N.A.

Annexure I (a) (Contd....)

Sl. No.	Name of MFP(EFP)	Botanical term	Area where available in abundance	Season of availability	Average annual collection in qtl.	Average procurement value per qtl. in Rs.	Sale value per qtl. in Rs. (1994-5)
7.	Cashew Nut	*Anacardium occidentale* Linn.	N.A.	March to April	N.A.	N.A.	N.A.
8.	Chakunda seeds	*Cassia tora* Linn.	Sambalpur, Khariar, Bolangir, Nowarangpur	N.A.	10768.66	240.00	N.A.
9.	Char Seeds	*Buchanania lanzan* spr.	Khariar, Rayagada Bolangir, Karanjia, Sambalpur	April to June	1061.58	300.00	N.A.
10.	Gaba seeds	N.A.	Baripada, Jeypore, Nowrangpur	N.A.	109.36	300.00	N.A.
11.	Gillo seeds	*Caesalpinia digyna* Rottl.	Mayurbhanj, Puri Keonjhar, Angul, Ganjam, Boudh	Nov. Dec	2595.56	175.00	N.A.
12.	Indrajaba seeds	*Holarrhena antidysenterica* Wall. ex. A.DC.	Rairakhol, Sambalpur, Sundergarh	N.A.	12.98	300.00	N.A.
13.	Jute seeds	N.A.	Nayagarh	N.A.	N.A.	N.A.	N.A.
14.	Karanja seeds	*Pongamia glabra* Vent.	Rayagada, Karanjia, Keonjhar, Parlakhemundi	Feb. to March	4987.07	450.00	N.A.
15.	Kurali seeds	N.A.	N.A.	N.A.	N.A.	N.A.	N.A.

16.	Kusum seeds	*Schleichera oleosa* (Lour.) Oken	Karanjia, Bonai, Keonjhar, Baripada, Jeypore	May to June	3921.22	610.00	N.A.
17.	Mahua seeds	*Madhuca indica* Gmel.	Jeypore, Baliguda, Phulbani.	April to May	30125.00	N.A.	N.A.
18.	Makhan seeds	*Canavalia gladiata* (Jacq.) DC	N.A.	N.A.	N.A.	N.A.	N.A.
19.	Mango kernel	*Mangifera indica* Linn.	Nowrangpur, Rayagada Jeypore	April to July	19772.28	325.00	N.A.
20.	Marda seeds	N.A.	Keonjhar, Bonai	N.A.	20124.34	1000.00	N.A.
21.	Nageswar seeds	*Mesua ferrea* Linn.	Mayurbhanj, Deogarh	N.A.	N.A.	N.A.	N.A.
22.	Neem seeds	*Azadirachta indica* A. Juss.	Bolangir	N.A.	2957.85	275.00	N.A.
23.	Palash seeds	*Butea monosperma* (Lam.) Taub.	Bolangir, Khariar	N.A.	50.16	N.A.	N.A.
24.	Polanga seeds	*Calophylium inophyllum* Linn.	Athamallik, Puri	N.A.	1.74	250.00	N.A.
25.	Sal Seeds	*Shorea robusta* Gaertn. f.	Jeypore, Puri Keonjhar, Nowrangpur	May to June	316801.43	175.00	275.00
26.	Salmaseeds	*Spondias pinnata* (L.f.) Kurz.	Baripada, Karanjia	January	N.A.	N.A.	N.A.
27.	Siali Seeds	*Bauhinia vahlii* Wt. & Arn.	Rayagada, Nayagarh, Nowrangpur	N.A.	N.A.	N.A.	N.A.
28.	Ritha	*Sapindus emarginatus* Wall.	Angul, Bolangir, Nowarangpur, Boudh	Feb. to March	93.52	600.00	N.A.

Annexure - I (a) (Contd....)

Sl. No.	*Name of MFP(EFP)*	*Botanical term*	*Area where available in abundance*	*Season of availability*	*Average annual collection in qtl.*	*Average procurement value per qtl. in Rs.*	*Sale value per qtl. in Rs. (1994-5)*
Edible Products							
1.	Arrow root	*Curcuma angustifolia* Roxb.	Nowrangpur, Jeypore	Feb. to April	1.28	4611.28	5257.00
2.	Mahua flower	*Madhuca indica* Gmel.	Jeypore, Baliguda, Phulbani.	Feb. to March	N.A.	N.A.	602.70
3.	Mushroom	N.A.	Keonjhar, Baripada, Karanjia.	July to October	267.85	200.00	680.00
4.	Tamarind	*Tamarindus indica* Linn.	Rayagada, Koraput, Keonjhar, Ganjam, Mayurbhanj	January to May	69479.11	257.98	353.55
Leaves							
1.	Kendu leaves	*Diospyros melanoxylon* Roxb.	Sundergarh, Patnagarh, Jeypore	April to May	487621.32	2L per 1 paise	2943.50
2.	Khajuri Pati	*Phoenix sylvestris* (L.) Roxb.	Keonjhar	N.A.	2100 qtls & 550 (bani)	2.00 per	4.00 per qtl.
3.	Palm leaves	*Palm*	N.A.	N.A.	N.A.	N.A.	N.A.
4.	Sal leaves	*Shorea robusta* Goertn. f.	Nayagarh, Keonjhar Jeypore	Oct. to Feb.	427847ch.	2.55 per ch.	4.57 per ch.
5.	Siali leaves	*Bauhinia vahlii* Wt. & Arn.	Nowrangpur, Rayagada Nayagarh	Dec. to Feb.	9857.10	153.50	287.80

6.	Versungha Leaves	*Murraya koenigii* (L.) Spreng.	N.A.	N.A.	309.00	50.00	N.A.
Animal, Mineral and Miscellaneous Products							
1.	Honey	—	Nowrangpur, Karanjia, Baripada, Rayagada, Jeypore	Feb to Aug	5.00	2600.00	3390.00
2.	Bee wax	—	Nowrangpur, Karanjia, Baripada, Rayagada, Jeypore	April to June	7.50	6000.00	6380.00
3.	Kusumi lac	*Schleichera oleosa* (Lour.) Oken	Bonai, Keonjhar, (Telkoi, Baripada)	Nov. to Feb.	248.00	1000.00	4000.00
4.	Soap Nut	*Sapindus emerginatus* Wall.	Ghumsur South, Rayagada	May to June	147.50	600.00	680.00
5.	Tamarind Seeds	*Tamarindus indica* Linn.	Rayagada, Koraput, Keonjhar, Bhawanipatna	April to May	49845.62	40.00	70.00
6.	Tassar cocoons	—	Bariapada, Karanjia	Nov.. to Dec	9790342Nos	0.85 P.C.	N.A.

Abbreviations

N.A.	Not Available	P.Ka	Per Kahan
P.K.	Per Kera	P.C.	Per a Cocoon
P.Ch.	Per Chaki	P.P.	Per Pana
P.B.	Per Bida		

Annexure II

Procurement and disposal of Kendu leaves by OFDC (Figures in Lakhs)

Crop year	*Quantity taken delivery in lakh qtl*	*Quantity disposed in lakh qtl*	*Sale value in lakh Rs.*	*Working Advance to CCF (KL) in lakh Rs.*	*Royalty to Govt. in lakh Rs.*	*Marketing Commission OFDC*
1989	4.55	4.55	15323.67	3059.13	11498.36	766.18
1990	3.96	3.93	11180.95	3912.36	6709.46	559.05
1991	4.96	4.95	13778.47	6181.09	6688.01	688.92
1992	5.08	5.08	13719.17	6713.14	6100.56	685.96
1993	4.96	4.94	14738.66	6330.98	7434.93	736.93
1994	4.91	3.81	11213.4	6552.02	4869.37	560.67

Annexure III(A)

Procurement and disposal of Sal seeds by OFDC from 1983-1994

Crop year	Collection charges per MT	Working expenses allowed by the Govt. inclusive of commission per MT	Cost of production per MT	Quantity collected in MT	Quantity sold in MT	Shortage in MT	Sale value realised (Rs. lakh)	Average selling price per MT	Actual working expenses except admn. cost Rs. lakh	Royalty paid to Govt. (Rs. lakh)	Profit (+) or Loss (–) (Rs. lakh)
1983	900	450	1350	45,136	44,728	408	785.57	1756.32	499.35	180.55	105.67
1984	900	450	1350	18,784	18,267	517	330.63	1810	178.44	83.99	68.20
1985	950	425	1375	23,685	22,889	796	503.57	2200	270.59	177.64	55.34
1986	1000	400	1400	33,363	32,357	1006	619.1	1913..35	420.57	171.39	27.14
1987	1000	400	1400	12,973	12,538	435	235.6	1879.04	168.12	62.09	5.39
1988	1100	400	1500	9,375	9,118	257	199.79	2191.1	138.49	70.39	8.09
1989	1100	400	1500	41,586	40,624	926	685.16	1686.59	585.85	69.05	30.26
1990	1500	500	2000	3,463	3,234	229	77.6	2399.35	67.84	11.32	1.56
1991	2060	650	2710	39,897	38,766	1131	1640.25	4231.12	975.91	420.5	243.84
1992	2500	770	3270	30,343	29,176	1167	681.49	2335.79	883.46	—	201.97
1993	1750	500	2250	35,162	34,315	847	731.29	2131.13	755.25	100	123.96
1994	1750	500	2250	23,499	22,843	656	621.77	2721.92	508.36		—Likely to be a profit year

Collection charges @ Rs.1500 per MT for 17,653 MT
Collection charges @ Rs. 2500 per MT for 22,334 MT
Accounts under finalisation

Annexure III(B)

Statement of expenditure for procurement of Sal seeds from 1987-94
(In Rupees per Metric Ton)

Sl. No.	Item of expenditure	1987 crop	1988 crop	1989 crop	1990 crop	1991 crop	1992 crop	1993 crop	1994 crop	1995 crop
1.	Transport	83.88	98.68	92.37	97.65	102.88	110.16	107.51	91.92	49.64
2.	Daily wages	19.89	21.50	12.01	33.59	27.58	20.28	17.55	12.85	12.91
3.	Establishment	43.38	20.47	13.09	13.68	4.41	11.18	9.36	3.72	8.93
4.	Godown rent	29.64	39.62	35.54	54.75	39.39	34.74	25.88	15.13	25.55
5.	Packing & Forwarding	15.45	16.73	19.95	22.33	39.96	35.69	37.55	27.11	35.53
6.	Insecticides	5.96	6.08	3.94	11.63	5.31	11.36	2.83	2.20	3.07
7.	Store and consumable	—	25.92	7.54	8.48	37.50	26.49	25.36	13.74	11.73
8.	Gunny bags	32.00	25.54	27.06	2.64	37.10	30.24	24.83	34.18	34.19
9.	Jeep hire charges & P.O.L.	19.20	23.82	11.57	24.46	11.40	11.55	11.62	11.42	13.64
10.	Miscellaneous	20.42	13.11	10.62	15.93	8.95	10.95	8.49	7.74	11.37
11.	Overhead	26.08	28.16	30.98	40.00	34.39	35.40	41.00	45.00	45.00
12.	Interest	—	57.68	44.20	134.00	37.20	73.53	86.00	111.71	111.71
		295.90	377.31	308.87	459.14	386.07	411.57	397.98	376.72	363.27
	Commission at Rs. 100 per MT	100.00	100.00	100.00	100.00	100.00	100.00	100.00	100.00	100.00
		395.90	477.31	408.77	559.14	486.07	511.57	497.98	476.72	463.27

Annexure III(C)

Divisionwise procurement of Sal seeds by OFDC from 1989-94

Sl. No.	*Name of Division*	*Quantity collected in Metric Tonnes*					
		1989	*1990*	*1991*	*1992*	*1993*	*1994*
1.	Bhubaneswar	31.466	—	25.57	11.234	4.004	14.570 E
2.	Angul (C)	858.152	9.857	436.121	658.79	491.78	159.416 D
3.	Angul (KL)	241.853	21.923	—	—	—	- E
4.	Dhenkanal	2490.9	160.2	2190.461	1057.614	400	1765.466 E
5.	Nayagarh	—	—	709.7625	—	80.322	- E
6.	Berhampur (C)	225.977	—	70.227	28.941	119.116	21.376 E
7.	Bhanjanagar (C)	1240.28	—	509.5805	826.853	27.196	330.868 D
8.	Munguda	676.614	—	917.2995	—	—	- E
9.	Bouch (C)	686.575	27.2675	412.9045	4234.222	234.222	- E
10.	Aska	198.235	—	239.0775	193.139	14.388	148.981 E
11.	Rayagada	—	—	298.153	120.4	219.125	28.732 E

12.	Sambalpur (C)	353.496	—	109.754	292.859	1299.47	10.261 D
13.	Jharsuguda (C)	4657.401	4.397	2827.439	1224.7595	3264.346	52.147 B
14.	Deogarh (C)	3892.428	101.319	1640.63	729.259	1532.873	1155.890 C
15.	Rairakhol	2138.235	26.436	678.4	218.13	728.639	27.650 D
16.	Rourkela	4952.121	10.832	1850.3675	1265.976	2539.919	2005.956 B
17.	Sundergarh	—	—	1746.527	1393.858	1688.765	176.828 D
18.	Bolangir (C)	494.113	48.28	261.824	251.48	186.767	322.520 E
19.	Bhawanipatna (C)	—	—	193.7595	86.918	45.462	—
20.	Jeypore (C)	4036.927	816.223	910.49	4153.952	1704.083	3262.985 A
21.	Nowrangpur	1441.681	1758.98	6308.633	12145.788	10871.77	7512.269 A
22.	Jajpur Road	—	99.037	3896.155	698.0835	1546.275	1920.108 C
23.	Keonjhar (C)	—	278.229	5484.749	2651.431	4197.758	3026.400 A
24.	Karanjia	—	21.59	4492.685	1008.992	1998.338	1867.582 B
25.	Baripada	—	78.19	2701.8495	1090.816	2161.151	1488.755 C
	Total	28616.454	3462.7605	38912.4190	34343.495	35355.769	25298.76

Annexure IV

Procurement and Sale value of EFP (other than Kendu leaves and Sal seeds) collected by OFDC from 1990-91 to 1995-6

Year	Name of the EFP items	Quantity collected in Qtl.	Sale value obtained in Rs.
1990-1	*Nux vomica*	3.30	990.00
	Honey	1.80	4290.00
	Resin	—	315.00
	Arrowroot	0.850	432.00
1991-2	*Nux vomica*	134.31	14905.00
	Siali leaf	44.87	13175.00
	Honey	21.330	73410.00
	Resin	32.690	7078.00
	Arrowroot	9.740	3494.00
	Sabai grass	110.00	—
	Harida	4.55	1308.00
	Bahada	175.74	1209.00
	Amla	4.88	1116.00
	Broom stick	933 bundles	2936.00
1992-3	*Nux vomica*	100.42	77735.00
	Siali leaf.	145.00	61625.00
	Honey	162.886	144313.00
	Resin	—	33249.00
	Arrowroot	11.284	40014.00
	Harida	77.36	—
	Bahada	436.65	—
	Amla	11.23	—
	Broom stick	85974 bundles	—

Annexure IV (Contd....)

Year	*Name of the EFP items*	*Quantity collected in Quintals*	*Sale value obtained in Rs.*
1993-4	*Nux vomica*	394.38	111206.00
	Siali leaf	111.00	49950.00
	Cane	190 (Pana)	15000.00
	K. Creeper	10.00	6000.00
	Hill-broom	75.00	48460.00
	Honey	203.445	1056911.00
	Arrowroot	2.00	13393.00
	Harida	5.94	213.00
	Bahada	5.84	144.00
	Amla	1.65	74.00
1994-5	*Nux vomica*	96.80	—
	Siali leaf	93.00	41950.00
	Hill broom	87.35	106600.00
	K. Creeper	5.00	3000.00
	Honey	10.47	284874.00
	Arrowroot	0.57	14772.00
	Resin	—	1327.00
1995-6	*Nux vomica*	—	—
	Siali leaf	97.00	46075.00
	Hill broom	16.00	12160.00
	K. Creeper	10.00	7600.00
	Honey	9.63	149545.00
	Arrowroot	—	6756.00

Annexure V

Yearwise procurement of Kendu leaves by KL division of Forest Department and expenditure chart during the period from 1984 to 1993

Year	*1984*		*1985*		*1986*		*1987*		*1988*	
Item	*Exp in lakh Rs.*	*Exp per qtl.*	*Exp in lakh Rs.*	*Exp per qtl.*	*Exp in lakh Rs.*	*Exp per qtl.*	*Exp in lakh Rs.*	*Exp per qtl.*	*Exp in lakh Rs.*	*Exp per qtl.*
Bush cutting	78.23	19.85	83.29	21.91	97.53	25.39	142.43	33.27	144.54	37.06
Repair & maintenance of phadi	104.36	26.48	112.11	29.50	129.02	33.59	167.38	39.110	152.80	39.17
Seasonal Staff	110.64	28.08	106.60	28.05	113.00	29.42	145.50	33.99	141.00	36.15
Misc. Contingencies	68.00	17.25	70.54	18.56	74.72	19.45	80.00	18.69	79.61	20.41
Estt. Cost	268.46	58.13	301.74	79.40	343.00	89.32	370.52	86.57	406.17	104.14
10% capital cost	28.32	7.18	32.55	8.56	36.24	9.43	39.55	9.24	42.91	11.00
Purchase price of K.L.	512.27	130.02	644.51	169.61	710.74	185.09	815.98	190.65	835.06	214.12
Drying and storage	85.10	21.60	105.29	27.71	113.62	29.59	160.07	37.40	137.04	35.14
Processing, binding & bagging	322.48	81.85	349.71	92.03	351.20	91.46	399.96	93.45	392.53	100.65
Transportation	38.37	9.74	38.65	10.10	40.66	10.59	49.43	11.55	41.88	10.74
Unforeseen	3.94	1.00	3.80	1.00	3.84	1.00	4.28	1.00	1.32	0.34
Research	—	—	—	—	—	—	—	—	7.60	1.95
World food programme	—	—	—	—	—	—	—	—	—	—
Total	1620.17	401.18	1848.82	486.51	2013.57	524.33	2375.10	554.91	2382.46	610.87

Annexure V (Cond..)

Year	1984		1985		1986		1987		1988	
Item	Exp in lakh Rs.	Exp per qtl.	Exp in lakh Rs.	Exp per qtl.	Exp in lakh Rs.	Exp per qtl.	Exp in lakh Rs.	Exp per qtl.	Exp in lakh Rs.	Exp per qtl.
Production in lakh qtl	3.94	—	3.80	—	3.84	—	4.28	—	3.90	—
Rate of purchase of KL from										
(a) Pluckers leaf/paisa	9L/1P	—	6L/1P	—	6L/1P	—	6L/1P	—	16L/3P	—
(b) Grower Pcs/Qtl.	125.00	—	150.00	—	150.00	—	150.00	—	160.00	—
Minimum wage as per L&E Deptt. per day	6.00	—	7.50	—	7.50	—	10.00	—	10.00	—
Total sale value received in lakh	3018.09	—	3672.75	—	3915.79	—	4319.39	—	5578.57	—
Average sale price in Rs.	779.00	—	998.00	—	1020.00	—	1048.00	—	1430.00	—
Royalty in lakhs Rs.	1271.72	—	1688.72	—	1812.15	—	1726.25	—	2921.60	—
OFDC Commission Rs. in lakhs	136.85	—	132.30	—	198.75	—	215.97	—	279.08	—
Bush cutting	165.18	36.30	163.52	41.08	383.75	77.21	380.00	74.65	383.09	76.92
Repair & maintenance of phadi	171.52	37.69	183.84	46.19	398.05	80.09	392.00	77.01	372.55	74.80
Seasonal staff	166.68	36.63	281.52	70.73	619.49	124.64	636.20	125.00	508.00	102.00

Misc. Contingencies	92.16	20.25	105.54	2651	151.08	30.39	180.21	35.40	166.50	32.93
Estt. Cost	437.58	96.17	497.00	124.87	526.58	105.95	617.39	121.29	745.70	149.73
10% capital cost	46.33	10.18	48.62	12.21	49.43	9.94	48.64	9.55	49.57	9.95
Purchase price of K.L.	1222.53	268.69	1738.78	436.88	2640.90	531.37	2857.01	561.30	2747.30	551.66
Drying and Storage	182.36	40.08	182.48	45.85	388.15	78.10	458.10	90.00	372.08	74.69
Processing, Binding & Bagging	485.66	106.74	644.44	161.92	907.02	182.50	992.55	195.00	909.00	182.53
Transportation	55.55	12.21	54.36	13.66	83.24	16.75	103.68	20.37	105.00	21.08
Unforeseen	2.91	0.64	0.99	0.25	5.21	1.05	12.26	2.41	10.00	2.00
Research	—	—	—	—	1.82	0.36	2.00	0.39	4.00	0.80
World food programme	19.47	4.28	16.67	4.19	26.34	5.30	33.08	6.50	27.00	5.42
Total	3047.93	669.86	3917.76	984.34	6181.06	1243.65	6713.12	1318.87	6400.00	1284.51
Production in lakh qtls.	4.55	—	3.98	—	4.97	—	5.09	—	4.93	—
Rate of purchase of KL from			up to15.590	after15.590						
(a) Pluckers leaf/paisa	14L/3P	4OL/9P	18L/5P	2L/1P	21/1P	—	2L/1P	—	2L/1P	—
(b) Grower Pcs/Qtl.	170.00	—	200.00	500.00	500.00	—	650.00	—	700.00	—
Minimum wage as per L&E Deptt. per day	11.00	—	11.00	25.00	—	25.00	25.00	—	25.00	—
Total sale value	15326.67	—	11180.95	—	13770.47	—	13716.81	—	14651.37	—

Annexure - V (Cond..)

Year	*1984*		*1985*		*1986*		*1987*		*1988*	
Item	*Exp in lakh Rs.*	*Exp per qtl.*	*Exp in lakh Rs.*	*Exp per qtl.*	*Exp in lakh Rs.*	*Exp per qtl.*	*Exp in lakh Rs.*	*Exp per qtl.*	*Exp in lakh Rs.*	*Exp per qtl.*
received in lakh									up to 31/10/94	
Average sale price in Rs.	3367.00	—	2809.00	—	2772.00	—	2695.00	—	2942.00	—
Royalty in lakhs Rs.	11498.36	—	6701.25	—	6893.12	—	6098.44	—	7284.31	—
OFDC commission Rs. in lakh	766.18	—	558.61	—	704.29	—	905.30	—	966.99	—
Bush cutting	165.18	36.30	163.52	41.08	383.75	77.21	380.00	74.65	383.09	76.92
Repair & maintenance of phadi	171.52	37.69	183.84	46.19	398.05	80.09	392.00	77.01	372.55	74.80
Seasonal staff	166.68	36.63	281.52	70.73	619.49	124.64	636.20	125.00	508.00	102.00
Misc. contingencies	92.16	20.25	105.54	2651	151.08	30.39	180.21	35.40	166.50	32.93
Estt. cost	437.58	96.17	497.00	124.87	526.58	105.95	617.39	121.29	745.70	149.73
10% capital cost	46.33	10.18	48.62	12.21	49.43	9.94	48.64	9.55	49.57	9.95
Purchase price of K.L.	1222.53	268.69	1738.78	436.88	2640.90	531.37	2857.01	561.30	2747.30	551.66
Drying and storage	182.36	40.08	182.48	45.85	388.15	78.10	458.10	90.00	372.08	74.69

Processing, binding & bagging	485.66	106.74	644.44	161.92	907.02	182.50	992.55	195.00	909.00	182.53
Transportation	55.55	12.21	54.36	13.66	83.24	16.75	103.68	20.37	105.00	21.08
Unforeseen	2.91	0.64	0.99	0.25	5.21	1.05	12.26	2.41	10.00	2.00
Research	—	—	—	—	1.82	0.36	2.00	0.39	4.00	0.80
World food programme	19.47	4.28	16.67	4.19	26.34	5.30	33.08	6.50	27.00	5.42
Total	3047.93	669.86	3917.76	984.34	6181.06	1243.65	6713.12	1318.87	6400.00	1284.51
Production in lakh qtls.	4.55	—	3.98	—	4.97	—	5.09	—	4.93	—
Rate of purchase of KL from			up to15.590	after15.590						
(a) Pluckers leaf/paisa	14L/3P	4OL/9P	18L/5P	2L/1P	21/1P	—	2L/1P	—	2L/1P	—
(b) Grower Pcs/Qtl	170.00	—	200.00	500.00	500.00	—	650.00	—	700.00	—
Minimum wage as per L&E Deptt. per day	11.00	—	11.00	25.00	—	25.00	25.00	--	25.00	—
Total sale value received in lakh	15326.67	—	11180.95	—	13770.47	—	13716.81	—	14651.37 up to 31/10/94	—
Average sale price in Rs.	3367.00	—	2809.00	—	2772.00	—	2695.00	—	2942.00	—
Royalty in lakh Rs.	11498.36	—	6701.25	—	6893.12	—	6098.44	—	7284.31	—
OFDC commission Rs. in lakh	766.18	—	558.61	—	704.29	—	905.30	—	966.99	—

Annexure VI

Divisionwise procurement of Kendu leaves and delivery to OFDC from 1990 to 1994

1990 - Crop

Name of KL Divn.	*Production (in qtl)*	*Delivery (in qtl)*	*Lifting (in qtl)*	*Balance to be lifted (in qtl)*
Angul	14,800.00	14,794.20	14,791.80	2.40
Athamallik	36,343.80	36,343.80	36,078.60	265.20
Keonjhar	21,855.15	21,842.40	21,831.60	10.80
Boudh	22,158.80	22,130.90	21,972.00	158.90
Phulbani	17322.60	17,322.60	13982.40	3340.20
Sambalpur	19,160.40	19,113.60	18,762.60	351.00
Rairakhol	22,042.20	21,985.20	21,805.20	180.00
Deogarh	22,101.40	22,085.50	22,085.40	.10
Jharsuguda	18,368.90	18,344.40	18,224.40	120.00
Rourkela	31,197.60	30,257.46	29,176.80	1080.66
Kuchinda	19,411.90	19,386.60	19,386.60	—
Bolangir	21,685.40	21675.00	21675.00	—
Khariar	15,661.51	15,657.40	15,523.85	133.55
Bhawanipatna	19,291.80	19,291.80	18,291.00	1000.80
Titlagarh	22,689.00	22,677.60	20,910.00	1767.60
Patnagarh	32,151.35	32,150.00	31,783.80	366.20
Padampur	27,564.60	27,564.60	27,504.60	60.00
Jeypore	6,416.35	6,416.35	6,237.68	178.67
Nowrangpur	7,179.55	6,971.25	6,971.25	—
	3,97,402.31	3,96,010.66	3,86,994.58	9,016.08

Annexure VI (contd...)

1990 - Crop

Name of KL Divn.	*Production (in qtl)*	*Delivery (in qtl)*	*Lifting (in qtl)*	*Balance to be lifted (in qtl)*
Angul	25,042.10	25,032.00	24,971.40	60.60
Athamallik	40,779.60	40,779.60	40,779.60	—
Keonjhar	23,261.40	23,261.40	23,261.40	—
Boudh	30,082.80	30,082.80	29,962.80	120.00
Phulbani	28,765.60	28,726.20	27,921.00	805.20
Sambalpur	24,836.80	24,829.80	24,829.80	—
Rairakhol	23,541.30	23,536.80	23,470.20	66.60
Deogarh	30,850.80	30,840.60	30,840.60	—
Jharsuguda	19,511.80	19,473.60	19,473.60	—
Rourkela	29,866.65	29,800.44	29,752.80	47.64
Kuchinda	23,534.45	23,481.31	23,481.31	—
Bolangir	23,409.60	23,257.20	23,196.60	60.60
Khariar	22,079.68	22,079.68	22,079.68	—
Bhawanipatna	22,534.46	22,484.66	22,484.66	—
Titlagarh	24,787.80	24,787.80	24,787.80	—
Patnagarh	36,279.30	36,186.00	36,186.00	—
Padampur	33,788.40	33,788.40	33,661.20	127.20
Jeypore	24,250.49	23,105.77	23,105.77	—
Nowrangpur	22,254.28	22,178.84	22,178.84	-
	5,09,457.31	5,07,712.90	5,06,425.06	1,287.84

Annexure - VI (contd...)

Figure of production of Kendu leaf 1993 - Crop
(Based on the RMs furnished by the DFOs)

Cuttack (KL) Circle

Name of KL Divn.	*Production (in qtl)*	*Delivery (in qtl)*	*Lifting (in Qtl)*	*Remark*
Angul	22,603.20	22,602.60	17,602.80	0.60 damaged
Athamallik	43,396.80	43,395.00	29,973.60	1.80 reasons not reported
Keonjhar	22,593.00	22,590.00	22,371.00	3.00 Theft
Boudh	27,802.80	27,802.80	27,802.80	—
Phulbani	24,674.40	24,661.20	20,171.60	13.20 Theft
Total	1,41,070.20	1,41,051.60	1,17,921.80	13.60

Sambalpur (KL) Circle

Name of KL Divn.	*Production (in qtl)*	*Delivery (in qtl)*	*Lifting (in qtl)*	*Remark*
Sambalpur	23,551.35	23,544.00	Not reported	6.15 Theft 1.20 Shortage 7.35 Total
Rairakhol	25,067.40	25,067.40	23,067.40	—
Deogarh	30,859.80	30,859.80	30,859.80	—
Jharsuguda	17,298.65	17,250.00	17,250.00	42.00 Theft 6.65 Burnt 48.65 Total
Rourkela	28,186.55	28,186.20	28,106.20	0.35 Loss
Kuchinda	19,583.55	19,455.60	19,455.60	123 Burnt 4.95 Theft 127.95 Total
Total	1,44,547.30	1,44,363.00	1,18,739.00	184.30

Annexure VI (contd...)

Bolangir (KL) Circle

Name of KL Divn.	*Production (in qtl)*	*Delivery (in qtl)*	*Lifting (in qtl)*	*Remark*
Bolangir	23,463.00	23,463.00	23,463.00	—
Khariar Proc	7,369.20	7,369.20	Not reported	—
Phal	13,993.86 PPQ	13,993.86 PPQ	—	—
Total	44,826.06	44,826.06	23,463.00	—
Bhawanipatna	22,624.80	22,599.00	22,399.00	19.80 Burnt 6.00 Theft 25.80 Total
Titlagarh	25,803.00	25,803.00	21,486.40	—
Patnagarh	35,269.00	35,269.00	35,269.00	—
Jeypore	26,473.66 PPQ	25,330.54 PPQ	24,447.74	137.68 damaged 1005.44 undelivered 1143.12 Total
Padampur	34,263.60	34,263.60	34,263.60	—
Nowrangpur	23,237.73	23,143.80	22,693.96	93.84 damaged
Total	1,67,671.79	1,66,408.94	1,60,559.70	1262.76
Grand Total	4,98,115.35	4,96,649.69	4,22,913.50	1465.66

Annexure VI (contd...)

Final Production - KL Operation
1994 Crop

Sl. No.	Name of Division	FinalProduction in Qntl/PPQ in case of Phal area			Padta
		P.L.	G.L.	Total	
1.	Angul	27,732.00	—	27,732.00	556.07
2.	Athamallik	26,355.00	10,452.6	36,807.60	613.14
3.	Keonjhar	21,936.00	—	21,936.00	529.80
4.	Boudh	27,031.20	—	27,031.20	583.11
5.	Phulbani	23,215.20	—	23,215.20	502.42
	Total	1,26,269.40	10,452.6	1,36.722.00	2784.54
6.	Sambalpur	23,766.25	928.20	24,694.45	607.81
7.	Rairakhol	25,427.40	3,042.60	28,470.00	646.40
8.	Deogarh	23,232.00	—	23,232.00	527.24
9.	Kuchinda	17,672.40	—	17,672.40	585.70
10.	Jharsuguda	16,708.80	—	16,708.80	578.45
11.	Rourkela	32,000.45	—	32,000.45	478.60
	Total	1,38,807.30	3,970.80	1,42,778.10	3,424.20
12.	Bolangir	23,124.60	—	23,124.60	525.34
13.	Patnagarh	33,517.20	—	33,517.20	519.53
14.	Titlagarh	26,128.20	—	26,128.20	533.38
15.	Padampur	30,067.80	—	30,067.80	582.33
16.	Bh. Patna	22,930.20	—	23,930.20	556.07
17.	Khariar (Pr. 6,712.80+PPQ 14,601.49)	21,314.29	—	21,314.29	688.38
18.	Jeypore	30,508.85 PPQ	—	30,508.85 PPQ	515.87
19.	Nowrangpur	24,459.45 PPQ	—	24,459.45 PPQ	566.92
	Total	2,12,050.59	—	2,12,050.59	4487.82
	Grand Total	4,77,127.29	14,423.40	4,91,550.69	10,696.56

Annexure VII

Itemwise procurement of EFP by TDCC during last 10 years
(Quantity in quintals)

Sl.	Items	1985-6	1986-7	1987-8	1988-9	1989-90	1990-1	1991-2	1992-3	1993-4	1994-5
1.	Seeded Tamarind	22,557.00	50,109.00	40,090.58	84,417.54	11,881.43	1,00,845.37	78,448.71	58,338.61	53,410.80	55,599.00
2.	De-seeded Tamarind	1,607.00	2,787.00	1,725.00	3,153.14	679.76	5,944.83	3,563.20	796.79	5,696.02	8,665.02
3.	Mahua seed	2,644.00	345.00	3,221.00	1,195.58	375.46	142.08	2,469.75	—	—	—
4.	Kusuma seed	281.00	2,230.00	254.02	23.60	591.06	672.135	17.74	—	—	—
5.	Neem seed	370.00	326.00	925.70	48.76	—	—	—	—	—	—
6.	Karanja seed	317.00	435.00	106.07	1,890.82	3.04	426.02	084.67	—	—	—
7.	Sal seed	10,611.86	48,958.00	16,561.00	1,68,756.87	—	2,194.55	—	19,997.19	40,704.87	6,228.00
8.	Harida	12,857.00	12,662.00	7,090.90	5,439.22	9,170.76	33,484.58	5,626.04	7,715.95	40,764.07	10,835.00
9.	Nux vomica	422.00	1,406.00	807.14	168.59	324.97	1,454.40	283.22	221.08	3,036.36	645.95
10.	Shikakai	30.00	44.00	204.98	272.05	154.03	18.91	—	—	—	—
11.	Hill-brooms (L)	1,563.00	2,251.00	2,231.20	1,538.00	1,369.10	1,560.75	854.23	848.97	2,255.34	2,419.00
12.	Hill-brooms (B)	—	10,530 nos	48,378 nos	67,677 nos	44,500 nos	85,983 nos	45,087 nos	40.67 nos	5,347 nos	1,16,383 nos
13.	Dry Siali leaves	1,427.00	1,055.00	1,163.77	1,700.15	1,403.17	2,512.12	2,925.71	5,771.60	9,265.36	10,105.00
14.	Stitched Siali leaves	5,469 (B)	12,456 (B)	861 (B)	987 (B)	4,112 (B)	7,520 (B)	—	—	698 (B)	—
15.	Kusumi lac	50.00	59.00	53.70	68.27	183.27	211.07	25.52	66.70	209.04	280.00

Annexure - VII (Contd...)

Sl.	Items	1985-6	1986-7	1987-8	1988-9	1989-90	1990-1	1991-2	1992-3	1993-4	1994-5
16.	Chargum	155.00	318.00	166.00	8.50	42.63	104.50	37.55	266.06	297.00	850.00
17.	Bahada	36.00	69.28	264.77	494.47	413.04	972.24	493.08	14.71	191.90	436.00
18.	Amla	51.00	46.00	13.12	42.12	7.49	528.27	42.28	11.98	13.97	29.65
19.	R.S. Root	—	—	5.55	—	—	1.06	—	—	8.50	12.50
20.	Sunari bark	645.66	498.00	186.68	666.32	10.00	101.25	—	477.52	90.40	790.00
21.	Sal leaf plates	46,763 ch	1,26,462 ch	1,56,864 ch	1,04,129 ch	1,10,968 ch	1,66,039 ch	2,12,211 ch	4,23,796 ch	6,17,856 ch	2,37.810 ch
22.	Siali leaf plates	2,51,190 ch	1,80,308 ch	2,67,968 ch	2,15,866 ch	2,06,292 ch	3,05,620 ch	97,827 ch	3,16,337 ch	4,31,010 ch	4,30,040 ch
23.	Dhaura gum	88.45	18.70	7.00	—	6.10	15.90	12.00	16.00	8.00	45.80
24.	Genduli gum	0.04	0.07	—	—	—	2.79	24.60	103.16	93.00	112.99
25.	Mahua flower	—	—	—	15,309.10	120.50	15,192.44	2,78,245.62	—	—	—
26.	Broom Grass	—	—	—	—	—	—	1,300.00	1992.00	3076.70	3,089.00
27.	Tamarind seeds	—	—	—	—	—	—	220.00	—	149.00	720.00
28.	Bahada gum	—	—	—	—	—	—	—	85.88	386.90	499.30
29.	Thorn broom	—	—	—	—	—	—	—	825.00	2387.50	1340.87
30.	Gum	—	—	—	—	—	—	—	739.00	419.10	710.00
31.	Mushorooms	—	—	—	—	—	—	—	18.00	141.71	394.90
32.	Kheluan Creeper	—	—	—	—	—	—	—	32250 kera	37350 kera	22,900 kera
33.	Mersinga leaves	—	—	—	—	—	—	—	180.00	619.00	—
34.	Soap nut	—	—	—	—	—	—	—	200.00	165.00	130.00
35.	Sabai grass	—	—	—	—	—	—	—	119.00	4000.00	5200 pana

Annexure VIII (A)

Procurement rate of EFP by TDCC during last 10 years (Average price per qtls.)

Sl.	Items	1985-6	1986-7	1987-8	1988-9	1989-90	1990-1	1991-2	1992-3	1993-4	1994-5
1.	Seeded Tamarind	168.00	119.00	164.00	200.20	220.00	222.10	192.85	225.00	227.60	253.28
2.	De-seeded Tamarind	283.00	205.00	255.00	300.00	302.35	334.15	325.95	325.00	415.55	462.21
3.	Myrobalan	60.00	54.00	58.00	89.00	143.95	142.15	144.80	150.00	132.05	199.08
4.	Mahua seed	316.00	356.00	639.50	470.80	469.50	479.25	501.25	—	—	—
5.	Kusum seed	150.00	191.00	324.00	257.35	234.10	327.80	299.85	—	—	—
6.	Karanj seed	124.00	145.00	226.00	214.40	200.00	257.40	256.50	—	—	—
7.	Neem seed	103.00	113.00	212.00	180.00	—	—	—	—	—	—
8.	Sal seed	100.00	100.00	110.00	110.00	—	150.00	—	250.00	175.00	175.00
9.	Genduli gum	900.00	1000.00	—	—	—	3068.00	2520.30	2500.00	2505.25	2690.50
10.	Bahada gum	—	425.00	952.00	500.00	—	518.85	483.80	600.00	697.85	1029.45
11.	Chargum	45.00	76.00	70.00	71.00	69.90	85.75	100.00	100.00	100.00	100.00
12.	Dhaura gum	500.00	596.00	650.00	—	599.20	577.25	628.50	600.00	635.00	676.86
13.	Nux vomica	101.00	121.00	132.00	181.35	181.00	214.30	218.60	298.75	662.95	558.87
14.	Kusumi lac	1832.00	975.00	892.00	816.15	747.50	824.85	709.35	1697.40	1851.30	1935.72
15.	Hill-brooms	266.00	287.00	289.00	299.20	226.25	354.40	354.74	369.30	392.90	433.24

Annexure - VII (Contd...)

Sl.	Items	1985-6	1986-7	1987-8	1988-9	1989-90	1990-1	1991-2	1992-3	1993-4	1994-5
16.	Shikakai	—	386.00	361.00	344.00	329.95	326.25	—	—	—	—
17.	Bahada	40.00	40.00	40.00	50.00	50.50	62.35	73.50	68.00	52.30	73.39
18.	Amla	60.00	60.00	65.00	73.55	86.20	164.35	168.00	—	150.30	168.63
19.	Sunari bark	50.00	50.00	50.00	59.16	60.00	64.95	—	150.00	143.80	126.54
20.	R.S. roots	600.00	—	550.00	—	—	700.00	—	—	1058.80	1520.00
21.	Sal resin	—	—	700.00	—	—	—	—	2500.00	—	—
22.	Siali leaves	50.00	67.00	76.00	75.20	33.60	110.30	116.10	91.13	141.05	153.49
23.	Siali leaf plates	1.60p.ch	1.20p.ch	1.40p.ch	1.65p.ch	2.00p.ch	2.50	2.50	2.56	3.00	3.29p.ch
24.	Sal leaf plates	1.20p.ch	1.20p.ch	1.30p.ch	1.45p.ch	2.00p.ch	2.00	2.00	2.00	2.5	2.50p.ch
25.	Mahua flower	—	—	—	150.75	—	3.00	—	—	—	—
26.	Broom grass	—	—	—	—	—	137.66	125.00	125.00	126.75	135.00
27.	Tamarind seeds	—	—	—	—	—	56.75	50.00	—	—	—
28.	Thorn broom	—	—	—	—	—	—	—	149.00	150.00	149.91
29.	Gum	—	—	—	—	—	—	—	612.00	651.50	487.33
30.	Mushorooms	—	—	—	—	—	—	—	187.50	197.60	200.00
31.	Kheluan creeper	—	—	—	—	—	—	—	0.95	0.98	1.79 per kahan
32.	Mersinga leaves	—	—	—	—	—	—	—	50.00	50.00	—
33.	Soap nut	—	—	—	—	—	—	—	600.00	600.00	600.00
34.	Sabai grass	—	—	—	—	—	—	—	42.00	50.00	3.00 per kahan

Annexure VIII(B)

Quantity and value of EFP procured by TDCC from 1992-3 to 1994-5

Sl. No.	Commodity	1992 to 1993		1993 to 1994		1994 to 1995	
		Quantity in qtl	*Value in lakh*	*Quantity in qtl*	*Value in lakh*	*Quantity in qtl*	*Value in lakh*
1.	Seeded Tamarind	58338.61	131.18	53410.80	121.56	55599.00	140.82
2.	De-seeded Tamarind	796.79	2.36	5696.02	23.67	8665.00	40.05
3.	Hill-Brooms	848.97	3.04	2555.34	10.14	2419.00	10.48
4.	Myrobalan (Harida)	7715.95	11.20	14874.40	19.33	10835.00	21.57
5.	Honey	3.32	0.10	4.96	0.20	3.18	0.14
6.	Siali leaves	5771.60	4.34	9265.36	13.23	10105.00	15.51
7.	Nux vomica	221.08	0.66	3036.36	20.13	735.15	4.17
8.	Marking nut	40.00	0.62	7.50	0.01	403.60	0.20
9.	Cleaning nut	—	—	50.62	0.25	51.67	0.26
10.	Mersinga leaves	240.00	0.12	619.00	0.31	NA	NA
11.	Antudi flower	—	—	76.50	0.08	44.00	0.04
12.	Dhataki flower	204.72	0.27	52.95	0.08	723.40	1.37
13.	Swan climber	—	—	272.00	0.11	363.00	0.16
14.	Sabai grass	278.40	0.14	278.00	0.14	5200 pana	0.19
15.	Thorn broom	1372.00	2.06	2387.50	3.58	1340.87	2.01

Annexure VIII (B) (Contd...)

Sl. No.	Commodity	1992 to 1993		1993 to 1994		1994 to 1995	
		Quantity in qtls	Value in lakhs	Quantity in qtls.	Value in lakhs	Quantity in qtls.	Value in lakhs
16.	Genduli gum	103.10	2.58	93.00	2.33	112.99	3.04
17.	Dhura gum	16.00	0.10	8.00	0.05	45.80	0.31
18.	Amata gum	44.00	2.50	1558.00	12.86	1387.00	11.32
19.	Bahada gum	249.87	2.02	385.00	2.70	499.30	5.14
20.	Char gum	266.06	0.27	297.00	0.29	850.00	0.85
21.	Other gum	735.00	4.55	419.10	2.73	710.00	3.46
22.	Lac	66.70	1.18	209.04	3.89	280.00	5.42
23.	Nageswar flower	—	—	1.50	0.02	4.75	0.07
24.	Sunari bark	477.52	0.72	90.40	0.13	790	3.46
25.	R.S. root	—	—	8.50	0.09	12.50	0.19
26.	Siali fibres	—	—	134.12	0.60	720.97	6.25
27.	Kathalai	—	—	17.70	0.24	16400 pana	0.21
28.	Resin	7.74	0.15	—	—	N.A.	N.A.
29.	Mushroom	18.00	0.11	141.70	0.28	394.90	0.79
30.	Bhaliakoli	—	—	210.00	0.17	304.00	0.25
31.	Fenfena fruit	—	—	25.40	0.04	8.90	0.02
32.	Simuli cotton	20.40	0.23	8.40	0.09	5.25	0.05

33.	Bhuin neem	—	—	200.00	0.10	256.66	0.13
34.	Cane	76 kahana	0.03	134 kahana	0.14	98 kahana	0.13
35.	Kheluan creeper	32250.00 kera	0.32	37350.00 kera	0.37	22900 kera	0.51
36.	Bahada	14.71	0.01	191.90	0.11	436.00	0.32
37.	Amla	11.98	0.2	13.97	0.21	29.65	0.05
38.	Sal leaf plates	423.796 chaki	11.61	617856 chaki	19.62	237810 ch	5.94
39.	Siali leaf plates	316337	032	431010	10.73	430040 ch	14.16
40.	Siali leaves processed	—	—	4344	0.47	8023.00	0.96
41.	Broom grass	1992.00	2.49	3076.70	3.90	3089.00	4.17
42.	Sal seeds	19897.19	49.74	40704.87	71.03	6288.00	11.00
43.	Soap nut	200.00	1.20	165.00	0.99	130.00	0.78
44.	Rohini flower	—	—	450.00	0.32	24.00	0.02
45.	Tamarind seed	—	—	149.00	0.06	720.00	0.29
46.	Khajuri pati	—	—	4200.00	0.08	1100 bari	0.08
47.	Pita alu	—	—	49.50	0.04	79.00	0.02
48.	Mankada kendu	—	—	—	—	23.00	0.02
49.	Chiranjee seed	—	—	—	—	16.00	0.07
50.	Antia fibre	—	—	—	—	44.00	0.04
51.	Sal processed crop	—	—	—	—	7830	0.16
	Total	**416288.714**	**268.10**	**580,749.21**	**347.5**	**124402.54**	**316.65**

Annexure IX

Districtwise procurement of EFP by TDCC of Orissa Ltd. during the year 1990-1 in quintals

Sl.	Commodity	Koraput		Kalahandi		Keonjhar		Mayurbhanj	
No.	MFP(EFP)	Qty.	Value	Qty.	Value	Qty.	Value	Qty.	Value
1.	Seeded Tamarind	62881.41	14069625.00	1265.16	284586.00	8792.92	1763250.50	7937.82	153156.00
2.	De-seeded Tamarind	6045.01	2026832.00	—	—	—	—	—	—
3.	Karanj seeds	—	—	—	—	—	—	—	—
4.	Dhaura gum	—	—	—	—	6.00	3600.00	7.46	4041.00
5.	Char gum	—	—	—	—	—	—	—	—
6.	Bahada gum	—	—	—	—	64.24	33489.00	—	—
7.	Shikakai	18.90	6169.50	—	—	—	—	—	—
8.	Mahua flower	8312.25	1988967.00	—	—	6486.77	1925874.00	1212.00	364450.00
9.	Hill-broom (L)	586.45	211169.00	432.65	159356.25	—	—	—	—
10.	Hill-broom (B)	28339 nos	49950.50	—	—	10686 nos	23778.00	—	—
11.	Siali leaf plates	—	—	—	—	—	—	—	—
12.	Sal leaf plate	—	—	—	—	1140 ch	24829.00	—	—
13.	Harida	10295.82	1304795.00	7870.18	1243404.00	3401.54	510073.25	6012.40	900910.00
14.	Bahada	—	—	167.02	8544.00	419.03	31502.00	20.00	1400.00
15.	Amla	—	—	360.18	56273.00	39.16	5874.00	2.96	444.00
16.	R.S. root	—	—	1.06	742.00	—	—	—	—

17. Kusumi lac	—	—	—	—	35.55	29786.50	—	—
18. Dry Siali leaves	415.00	49752.00	—	—	—	—	—	—
19. Stitched Siali leaves	7632 (B)	37410.00	—	—	—	—	—	—
20. Mango kernel	—	—	—	—	—	—	—	—
21. Simuli cotton	—	—	—	—	—	—	2.50	3750.00
22. Kusum seeds	—	—	—	—	36.75	11025.00	—	—
23. Mohua seeds	—	—	—	—	—	—	16.50	10725.00
24. Sunari bark	—	—	—	—	—	—	—	—
25. Arrowroot	—	—	13.20	39600.00	—	—	5.00	1700.00
26. Nux vomica	—	—	5.30	1060.00	4.76	1071.00	811.85	183033.50
27. Honey	—	—	—	—	—	—	1.26	3528.00
28. Marking nut	—	—	—	—	638.25	35740.10	105.50	9930.00
29. Mushroom	—	—	—	—	—	—	95.00	28500.00
30. Sal seeds	—	—	1280.54	192081.00	—	—	—	—

Annexure IX (Contd....)

Sl. No.	Commodity MFP(EFP)		Koraput Qty.	Koraput Value	Kalahandi Qty.		Kalahandi Value	Keonjhar Qty.	Keonjhar Value	Mayurbhanj Qty.	Mayurbhanj Value
1.	Seeded Tamarind	2816.95	579000	18191.7	4228973.9	—	—	1202.48	270558	103088.38	22732778
2.	De-seeded Tamarind	—	—	—	—	—	—	—	—	6045.01	2026832.8
3.	Karanj seeds	7	1750	419.02	109242	—	—	—	—	426.02	110992
4.	Dhaura gum	7.12	4487	—	—	—	—	—	—	20.58	12128
5.	Char gum	119.16	10684.2	—	—	—	—	—	—	119.16	10684.2
6.	Bahada gum	13.41	6810.75	—	—	—	—	—	—	77.65	40299.75
7.	Shikakai	—	—	—	—	—	—	—	—	18.9	6169.5
8.	Mahua flower	230	60980	82.1	15712.5	—	—	—	—	16323.12	4355983.55
9.	Hill-broom (L)	—	—	563.82	186586	—	—	—	—	1573.92	557111.25
10.	Hill-broom (B)	—	—	52355 Nos.	78540	530 Nos.	874.5	—	—	91900 Nos.	153143
11.	Siali leaf plates	—	—	—	—	—	—	151088 ch.	422263	151088 ch.	422263
12.	Sal leaf plate	—	—	—	—	—	—	278456 ch.	555455.25	279596 ch.	580284.25
13.	Harida	4784.79	695578.5	880.46	117603.85	—	—	217.68	32336.75	33462.87	4804701.2
14.	Bahada	179.58	9129.25	51.74	3261	—	—	28.07	2154.75	865.44	55991
15.	Amla	25	2625	—	—	—	—	77.2	15130	504.5	80346
16.	R.S. root	—	—	—	—	—	—	—	—	1.06	742

17. Kusumi lac	176.48	144315.55	—	—	—	—	—	—	212.03	174102.05
18. Dry Siali leaves	2062.83	223931	—	—	—	—	—	—	2477.83	273683
19. Stitched Siali leaves	—	—	—	—	—	—	—	—	7632 (B)	37410
20. Mango kernel	—	—	726.23	99976.4	—	—	—	—	726.23	99976.4
21. Simuli cotton	—	—	—	—	—	—	—	—	2.5	3750
22. Kusum seeds	833.285	273420.5	—	—	—	—	—	—	870.035	284445
23. Mahua seeds	—	—	135.58	57367.7	—	—	—	—	142.08	68092.7
24. Sunari bark	—	—	—	—	—	—	101.25	6576.25	101.25	6576.25
25. Arrowroot	—	—	—	—	—	—	—	—	18.2	41300
26. Nux vomica	—	—	—	—	—	—	580.11	116222	1402.02	301386
27. Honey	—	—	1.52	4578	—	—	—	—	2.78	8103
28. Marking nut	—	—	—	—	—	—	—	—	803.75	45670.1
29. Mushroom	—	—	—	—	—	—	—	—	95	28500
30. Sal seeds	62.21	9331.5	—	—	733.51	110027.25	118.29	17743.50	2194.55	329183.25
Total									**171574.865qtl.**	**37652627.25**
									91900 nos.	
									430684 ch	

Annexure X

EFP items to be collected by TDCC

from different divisions in 1995-96

Sl. No.	Name of forest division	Name of forest range		Name of MFP(EFP) items to be operated
1.	Nayagarh	Entire Division	1.	Nux vomica
		"	2.	Siali leaves, plate and creeper
		"	3.	Sal leaves and plates
		"	4.	Tamarind
		"	5.	Keruan Creeper
		"	6.	Canes
		"	7.	Myrobolans
		"	8.	Sabai grass
		"	9.	Honey
2.	Ghumsar South	Entire Division	1.	Soap nuts
		"	2.	Tamarind
		"	3.	Hill-brooms
		"	4.	Keruan Creeper
		"	5.	Canes
		"	6.	Mersinga leaves
		"	7.	Nux vomica
3.	Puri	Balugaon Range	1.	Keruan Creeper
		"	2.	Hill-brooms
		Balugaon, Ranpur	1.	Honey and Wax
		& Tangi range	2.	Thorn Brooms
		Entire Division	3.	Nux vomica
		"	4.	Genduli gum
		"	5.	Tamarind
4.	Nowrangpur	Entire Division	1.	Tamarind
		"	2.	Hill-brooms
		"	3.	Siali leaves
		"	4.	Genduli and other

Annexure X (Contd...)

Sl. No.	*Name of forest division*	*Name of forest range*		*Name of MFP(EFP) items to be operated*
				gums
		"	5.	Arrowroots
		"	6.	Putrani roots
		"	7.	Dhatuki flower
		"	8.	Marking nuts
		"	9.	Bahada fruits
		"	10.	Amla
		"	11.	Harida
		"	12.	Siali leaves and fibres
		"	13.	Tamarind seeds
5.	Jeypore	Entire Division	1.	Tamarind
		"	2.	Hill-brooms
		"	3.	Genduli gum and other gums
		"	4.	Myrobalans
		"	5.	Arrowroots
		"	6.	Honey and Wax
		"	7.	Tamarind seeds
6.	Bolangir	Entire Division	1.	Amla
		"	2.	Harida
		"	3.	Bahada
		"	4.	Genduli gum
		"	5.	Thorn brooms
		"	6.	Tamarind
7.	Kalahandi	Entire Division	1.	Hill brooms
		"	2.	Tamarind
		"	3.	Harida
		"	4.	Bahada
		"	5.	Amla
		"	6.	Siali leaves
		"	7.	Genduli and other gums
8.	Khariar	Entire Division	1.	Myrobalans
		"	2.	Hill-brooms
		"	3.	Tamarind
		"	4.	Genduli and other gums
		"	5.	Thorn brooms

Annexure - X (Contd...)

Sl. No.	Name of forest division	Name of forest range	Name of MFP(EFP) items to be operated
		"	6. Dhataki flowers
		"	7. Marking nuts
		"	8. Siali fibres
		"	9. Siali leaves
9.	Rayagada	Muniguda, Gunupur & Gudari ranges	1. Tamarind
			2. Hill-brooms
		"	3. Myrobalans
		"	4. Nux vomica
		"	5. Siali leaves
		"	6. Marking and Cleaning nuts
		"	7. Honey and wax
		"	8. Genduli and other gums
		"	9. Soapnuts
		"	10. R.S. Roots
		"	11. Siali fibres
		Narayan Patna range	1. Tamarind
		"	2. Hill-brooms
		"	3. Nux vomica
		"	4. Siali leaves
		"	5. Genduli and other gums
		"	6. R.S. Roots
		"	7. Tamarind seeds
		Koraput and Lamtaput Range	1. Siali leaves
			2. Genduli and other gums
		"	3. R.S. Roots
		Kashipur and Rayagada range	1. Tamarind
			2. Hill-brooms
		"	3. Myrobalans
		"	4. Nux vomica
		"	5. Siali leaves
		"	6. Marking and cleaning nuts
		"	7. Genduli and other gums
		"	8. R.S. Roots

Annexure X (Contd...)

Sl. No.	Name of forest division	Name of forest range		Name of MFP(EFP) items to be operated
		"	9.	Siali fibres
10.	Parlakhemundi	Entire Division	1.	Tamarind
		"	2.	Hill-brooms
		"	3.	Harida
		"	4.	Bhalia fruits
		"	5.	Sabai grass
		"	6.	Nux vomica
		"	7.	Siali leaves
		"	8.	Kathalai
		"	9.	Honey
		R. Udayagiri, Mohana, Chandragiri range	10.	Genduli and other gums
11.	Sundergarh	Bisra and Kuarmunda	1.	Gum
		ranges	2.	Tamarind
		"	3.	Myrobalan
		"	4.	Lac
		"	5.	Siali leaves
		"	6.	Siali fibres
		In all other ranges	1.	Gum
		"	2.	Tamarind
		"	3.	Myrobalan
		"	4.	Lac
		"	5.	Siali leaves
			6.	Siali fibres
			7.	Broom grass
12.	Bamra	Entire Division	1.	Tamarind
		"	2.	Myrobalan
		"	3.	Siali leaves
		"	4.	Broom grass
		"	5.	Broom sticks
13.	Bonai	Entire Division	1.	Gum
		"	2.	Tamarind
		"	3.	Myrobalans
		"	4.	Lac
		"	5.	Siali leaves
		"	6.	Kanta bandhuni
		"	7.	Katakal
14.	Deogarh	Entire Division	1.	Tamarind

Annexure X (Contd...)

Sl. No.	Name of forest division	Name of forest range	Name of MFP(EFP) items to be operated
		”	2. Myrobalans
		Entire Division	3. Lac
		”	4. Siali leaves
		”	5. Broom grass
		”	6. Hill-brooms
		”	7. Sunari bark
		”	8. Nageswar flower
15.	Rairakhol	Entire Division	1. Tamarind
		”	2. Dhataki flower
		”	3. Myrobalans
		”	4. Sabai grass
		”	5. Nux vomica
		”	6. Chereita
		”	7. Siali leaves
		”	8. Genduli and other gums
		”	9. Lac
		”	10. Simuli cotton
16.	Keonjhar	Entire Division	1. Tamarind
		”	2. Lac
		”	3. Myrobalans
			4. Genduli and other gums
		”	5. Nux vomica
		”	6. Mushrooms
		”	7. Khajuri patia
		”	8. Siali leaves
		”	9. Atundi fruits
		”	10. Fenfena fruit
		”	11. Forest Bela
		”	12. Eksira fruit
		”	13. Cleaning nuts
		”	14. Marking nuts
17.	Karanjia	Entire Division	1. Tamarind
		”	2. Myrobalans
		”	3. Wax
		”	4. Gums
		”	5. Nux vomica

Annexure X (Contd...)

Sl. No.	Name of Forest Division	Name of Forest Range		Name of MFP(EFP) items to be operated
		"	6.	Mushrooms
		Entire Division	7.	Pita aloo fruits and flowers
		"	8.	Mankadakenda fruits
		"	9.	Swan climber
		"	10.	China climber
		"	11.	Chireita
		"	12.	Mushrooms
		"	13.	Marking nuts
		"	14.	Salma seeds
		"	15.	Punasi flower
		"	16.	Asana fruits
		"	17.	Ranja seeds
18.	Baripada	Entire Division	1.	Tamarind
		"	2.	Myrobalans
		"	3.	Lac
		"	4.	Gums
		"	5.	Nux vomica
		"	6.	Simuli cotton
		"	7.	Pita aloo fruits and flowers
		"	8.	Mankadakenda fruits
		"	9.	Swan climber
		"	10.	China climber
		"	11.	Chireita
		"	12.	Dhataki flower
		"	13.	Marking nuts
		"	14.	Salma fruits and seeds
		"	15.	Punasi flower
		"	16.	Asana fruits
		"	17.	Runja seeds
		"	18.	Rohini flower
		"	19.	Fenfena fruits and seeds
		"	20.	Katak seeds
		"	21.	Atundi fruits

Annexure XI(A)

Statement showing the quantity of EFP procured and the quantity processed by Utkal Forest Products Ltd. during 1992-95

Items	*Collection in Quintal*			*Processing in Quintal*		
Seeds	*1992-3*	*1993-4*	*1994-5*	*1992-3*	*1993-4*	*1994-5*
Mahua	12197.50	54127.38	6121.31	11113.13	50349.70	3788.29
Banatulsi	984.62	9676.71	5178.70	18736.21	13200.00	—
Chakunda	7563.81	5086.57	16449.75	—	6551.22	—
Gaba	40.59	111.22	107.51	—	26.20	164.45
Banakulthi	649.29	933.11	27.82	319.07	—	1739.63
Sikakai	62.25	121.88	65.16	—	—	—
Baghanakhi	21.05	11.07	01.25	—	—	—
Indrajaba	77.99	15.25	10.71	—	—	—
Banahaldi	01.05	01.27	0.47	—	—	—
Kamalagundi	29.98	30.97	20.15	—	—	123.30
Marda	105.03	10.42	0.76	—	—	131.05

Palash	152.68	100.36	34.09	—	—	249.27
Benacher	20.70	45.37	10.89	—	—	—
Babool	2162.56	2729.65	4047.07	962.80	2124.75	5731.00
Mango kernel	9646.15	1751.55	37793.01	7581.35	1564.73	—
Charseed	1026.70	486.66	1636.50	—	850.00	460.71
Gillo	1470.08	2775.11	2416.01	207.39	—	256.00
Anato	63.25	02.29	04.98	—	—	—
Baidanka	00.49	23.01	1.33	—	—	—
Karanja	730.50	7431.90	2542.25	515.72	5963.01	994.35
Neem	6118.99	2981.35	2934.35	4758.44	2890.39	—
Kusum	15271.09	2832.23	5010.22	5395.08	10614.31	—
Polang	116.10	—	0.74	—	—	—
Nageswar	01.30	—	—	—	—	62.71
Total	**58513.75**	**91285.33**	**84415.03**	**49589.19**	**94134.31**	**13700.76**

Annexure XI (B)

Statement showing processing cost, total collection, and sale value of EFP collected by Utkal Forest Products Limited

during 1992-93 to 1994-95

	1992-3 (*Rs.*)	*1993-4* (*Rs.*)	*1994-5* (*Rs.*)
Processing costs	44,56,960.20	1,18,15,558.00	11,06,082.53
Total collection costs	50,68,518.40	4,14,39,459.87	8,60,71,667.32
Total sales value	2,89,16,894.53	5,87,81,381.81	13,19,01,822.27

Annexure XI (C)

Divisionwise collection of EFP items processed by Utkal Forest Products Ltd. from 1.10.92 to 30.9.93 (In quintals)

Neem Seed	Char seed	Mahua seed	Chakunda seed	Karanja seed	Babool seed	Shikakai seed
1.24	13.22	1199.97	316.04	0.75	1.29	6.00
—	—	—	360.00	—	—	—
—	16.99	83.67	737.68	64.13	—	1.67
—	10.99	42.65	228.35	—	—	—
—	9.39	43.45	102.58	4.70	—	55.10
—	6.85	325.48	144.35	—	—	3.13
32.97	18.98	6357.41	38.72	38.38	6.43	—
—	16.13	8126.21	66.44	86.07	3.25	27.06
—	36.09	5716.72	30.17	855.41	0.40	0.95
—	17.40	3257.99	188.02	90.89	—	10.62
—	88.68	5493.07	635.10	36.51	—	—
—	0.67	332.27	515.99	48.32	—	—
—	—	3161.21	24.00	1342.23	—	0.40

Annexure XI(C) (Contd....)

Neem Seed	*Char seed*	*Mahua seed*	*Chakunda seed*	*Karanja seed*	*Babool seed*	*Shikakai seed*
467.96	—	179.80	673.90	21.30	1662.70	—
2041.07	—	52.49	1198.00	18.68	212.58	7.03
1.74	14.76	66.95	1367.04	0.39	626.05	—
4.40	6.29	494.06	241.45	264.40	—	—
58.10	—	5084.95	29.17	20.48	3.00	—
—	—	8508.78	66.00	13.27	—	—
48.28	16.56	84.66	150.54	2.03	—	—
—	—	3468.89	40.37	4314.65	—	9.43
17.70	—	54.00	15.00	3.57	—	—
—	—	—	1.40	17.50	—	—
—	—	347.84	13.31	3.33	—	—
0.07	6.68	20.55	5.34	13.62	—	
—	—	—	—	—	—	—
332.30	7.31	3.54	16.00	0.43	—	0.02
3005.83	**286.99**	**52506.61**	**7204.96**	**7261.04**	**2515.70**	**121.41**

2.62	1.18	0.38	14.66	503.16	27.30	1.38	—
—	—	2.00	20.00	—	—	—	—
40.58	—	0.19	0.10	38.60	57.63	64.79	—
10.00	1.06	—	4.00	146.95	0.10	7.38	—
—	—	—	6.00	305.45	4.20	2.42	—
65.44	—	0.08	44.24	44.69	—	13.68	—
4.08	—	0.14	—	463.96	—	1652.79	—
216.75	—	0.11	166.25	113.78	25.40	25.24	—
2.56	7.93	5.28	1.00	150.82	20.36	204.36	—
64.33	0.02	0.34	8.34	16.72	1.03	91.53	—
345.41	—	—	559.77	—	343.19	532.43	40.00
3.22	—	—	30.75	—	45.90	704.70	—
45.20	00	0.80	5.00	8.00	37.70	150.00	24.00
21.40	—	—	28.06	7.15	—	2.04	—
32.80	—	0.35	5.00	—	1.95	8.70	—

Annexure XI(C) (Contd....)

	Neem seed	*Char seed*	*Mahua seed*	*Chakunda seed*	*Karanja seed*	*Babool seed*	*Shikakai seed*
—	—	—	1.01	—	—	37.80	—
45.00	—	—	1.00	20.00	1.03	440.64	—
6.30	—	—	0.78	—	261.82	—	—
7.06	—	0.37	12.44	47.49	737.82	11.52	—
—	—	—	—	124.20	8.16	46.57	—
0.68	—	—	—	—	45.48	17.59	—
—	—	—	2.00	76.50	4.19	1.36	—
—	—	—	—	265.61	30.80	—	—
—	—	—	0.39	23.05	2.80	—	—
—	—	—	—	182.47	3.21	—	—
—	—	—	—	256.60	—	—	—
0.93	0.31	0.09	—	22.31	1.45	3.40	—
914.36	**10.50**	**10.13**	**910.79**	**2817.51**	**1661.52**	**4020.32**	**64.00**

Annexure - XI (D)

Divisionwise collection of EFP items by Utkal Forest Products Ltd.

for the period ending 30.9.94 (in quintals)

Sl. No.	Division	Anato	Gillo	Karanja	Banatulsi	Banakulthi	Baghanakhi	Char
1.	Baripada	—	229.01	44.33	16.00	—	—	11.75
2.	Karanjia	0.11	95.65	442.94	114.50	0.50	—	69.83
3.	Keonjhar	0.03	138.16	111.54	575.40	61.30	—	7.80
4.	Angul	—	227.46	1.36	—	—	—	—
5.	Dhenkanal	—	307.72	1.64	—	—	—	0.46
6.	Athamallik	—	40.97	2.69	—	—	—	0.27
7.	Sambalpur	—	23.98	0.04	43.33	0.10	0.16	127.46
8.	Rairakhol	—	90.64	—	28.01	—	—	6.97
9.	Deogarh	—	53.52	1.46	199.90	12.72	—	0.56
10.	Bamra	—	33.48	3.53	563.78	—	0.05	2.26
11.	Baliguda	—	21.75	—	—	—	—	—
12.	Phulbani	1.54	23.25	11.67	99.45	12.00	1.52	0.16
13.	Boudh	—	192.96	0.38	—	—	—	3.76

Sl. No.	Division	Anato	Gillo	Karanja	Banatulsi	Banakulthi	Baghanakhi	Char
14.	Parlakhemundi	—	—	404.31	—	—	—	—
15.	Nayagarh	—	86.79	4.80	—	—	—	1.80
16.	Puri	—	162.59	—	—	—	—	—
17.	Ghumsar (S)	—	49.76	2.66	—	—	—	—
18.	Ghumsar (N)	—	176.15	—	—	—	—	1.05
19.	Athgarh	—	163.73	0.11	—	—	—	—
20.	Sundergarh	—	0.32	59.67	629.75	—	—	1.91
21.	Bonai	—	11.34	77.99	88.33	6.70	1.50	3.02
22.	Jeypore	2.40	—	10.61	2568.08	241.93	—	5.43
23.	Nowrangpur	2.29	—	20.43	217.96	126.47	—	—
24.	Rayagada	—	8.68	1000.80	992.20	—	—	4.00
25.	Kalahandi	—	—	11.20	1469.32	218.70	1.50	0.90
26.	Bolangir	—	—	1.80	2252.45	—	0.37	—
27.	Khariar	—	—	—	—	—	—	1613.85
	Grand Total	**6.37**	**2137.51**	**2215.96**	**9858.46**	**680.42**	**5.10**	**1863.24**

1. Baripada	205.77	—	1516.21	60.23	2.80	0.95	—	—	—
2. Karanjia	291.57	—	1673.74	—	4.85	0.40	—	—	—
3. Keonjhar	376.33	—	886.99	—	23.60	—	0.37	—	—
4. Angul	—	—	0.38	—	3.32	—	—	—	—
5. Dhenkanal	164.69	—	0.49	0.72	3.08	—	—	—	—
6. Athamallik	1.30	0.74	2.85	0.25	44.74	100.00	—	—	—
7. Sambalpur	16.99	—	1.63	126.43	13.16	0.30	—	—	—
8. Rairakhol	3.62	—	0.16	—	24.34	—	—	—	—
9. Deogarh	50.82	—	16.21	0.04	37.46	—	—	—	—
10. Bamra	49.43	—	15.69	—	135.52	—	—	—	—
11. Baliguda	451.15	—	—	—	7.70	—	—	—	—
12. Phulbani	545.62	—	45.75	0.57	37.93	—	—	0.79	—
13. Boudh	—	—	—	52.24	5.60	—	—	—	—
14. Parlakhemundi	332.24	—	69.60	—	0.90	—	—	—	—
15. Nayagarh	—	—	—	—	—	—	—	—	—
16. Puri	—	—	—	—	—	—	—	—	—
17. Ghumsar (S)	95.60	0.70	—	—	—	—	—	—	—

Annexure XI(D) (Contd...)

Sl. No.	*Division*	*Mahua (S)*	*Polang*	*Kusum*	*Neem*	*Chakunda*	*Babool*	*Marda*	*Kamala gundi*	*Baidank*
18.	Ghumsar (N)	5.89	—	12.00	0.46	0.85	—	—	—	—
19.	Athgarh	—	—	—	3.25	—	—	—	—	—
20.	Sundergarh	246.72	—	235.40	1.05	115.75	—	—	0.10	—
21.	Bonai	925.84	—	244.51	0.08	86.50	—	0.67	3.58	—
22.	Jeypore	2476.83	—	39.40	00	74.56	—	—	—	—
23.	Nowrangpur	83.48	—	13.60	00	899.73	—	—	—	—
24.	Rayagada	1022.74	—	48.36	—	7.00	—	—	—	—
25.	Kalahandi	144.92	—	4.60	1204.75	288.70	2631.36	—	0.40	0.83
26.	Bolangir	42.03	—	74.64	1601.98	480.45	1042.26	—	14.58	0.50
27.	Khariar	63.03	—	4.35	—	1429.43	788.93	—	0.70	—
Grand	**Total**	**7596.61**	**1.44**	**4906.56**	**3052.05**	**3727.97**	**4564.20**	**1.04**	**20.15**	**1.33**

Annexure XI(E)

Divisionwise collection of EFP items by Utkal Forest Products Ltd.

from 1.10.94 to 30.9.95 (in quintals)

Sl. No.	Division	Palash	Gaba	Char	Benacher	Karanja	Gillo	Shikakai	Mahua
1.	Angul	—	—	—	—	3.01	272.26	13.34	—
2.	Athamalik	—	—	—	—	4.20	37.55	—	0.80
3.	Athagarh	—	—	—	—	0.71	202.38	—	—
4.	Keonjhar	0.06	—	—	—	330.84	161.29	3.20	4630.91
5.	Karanjia	—	—	—	—	3176.99	193.87	—	1852.90
6.	Jeypore	—	49.83	—	—	7.53	—	—	8514.11
7.	Deogarh	—	—	0.46	—	—	84.16	—	116.41
8.	Nayagarh	—	—	0.10	—	5.96	122.08	—	—
9.	Baripada	—	—	12.10	—	540.50	428.84	—	3703.83
10.	Dhenkanal	—	—	5.56	—	0.09	348.61	—	287.16
11.	Ghumsar (S)	—	—	5.21	—	2.09	169.45	—	401.11
12.	Ghumsar (N)	—	—	—	—	6.46	235.27	—	—
13.	Khariar	0.25	—	344.09	—	0.06	—	—	23.44

Annexure XI (E) (Contd....)

Sl. No.	Division	Palash	Gaba	Char	Benacher	Karanja	Gillo	Shikakai	Mahua
14.	Bolangir	20.86	0.35	143.62	—	4.51	—	3.62	90.40
15.	Bamra	—	—	0.80	—	2.62	21.02	0.88	38.27
16.	Baliguda	—	—	—	—	45.73	28.12	—	7935.96
17.	Rayagada	—	—	—	—	1755.02	4.10	—	5296.03
18.	Kalahandi	—	—	—	—	—	0.35	—	685.84
19.	Phulbani	—	—	0.94	—	33.05	31.54	0.28	7835.52
20.	Sundergarh	0.69	—	—	0.05	55.97	10.74	—	578.58
21.	Sambalpur	—	—	—	10.19	—	8.22	—	0.52
22.	Nowrangpur	—	19.66	—	—	23.30	—	—	840.50
23.	Boudh	—	—	—	0.13	2.00	131.66	0.18	35.93
24.	Bonai	—	—	1.68	—	189.68	7.56	0.16	1782.48
25.	Puri	—	—	—	—	—	260.66	—	—
26.	Rairakhol	—	—	—	—	—	127.39	—	—
27.	Parlakhemundi	—	—	—	—	4291.84	—	—	4983.61
	Grand Total	**21.86**	**69.84**	**514.56**	**10.37**	**10482.16**	**2887.12**	**21.66**	**49634.31**

1.	Angul	—	—	—	—	—	—	—	—
2.	Athamalik	—	—	—	—	—	—	—	—
3.	Athagarh	—	—	—	0.97	—	—	—	—
4.	Keonjhar	11.83	—	—	—	50.84	25.43	—	—
5.	Karanjia	24.70	—	—	—	15.25	26.37	—	—
6.	Jeypore	40.45	3.27	—	—	513.60	44.94	—	—
7.	Deogarh	—	—	—	0.36	15.17	2.29	—	—
8.	Nayagarh	1.00	—	—	—	—	—	—	—
9.	Baripada	4.28	—	—	11.08	26.65	38.42	—	—
10.	Dhenkanal	127.10	1.73	—	—	—	—	—	—
11.	Ghumsar (S)	—	—	—	—	—	0.90	—	—
12.	Ghumsar (N)	—	—	—	—	—	—	—	—
13.	Khariar	789.93	—	—	93.16	110.59	8.07	0.50	—
14.	Bolangir	209.69	4.96	0.15	1075.50	57.47	3.94	—	—
15.	Bamra	96.94	—	—	—	119.61	12.14	—	—
16.	Baliguda	7.00	—	—	—	24.40	—	—	—

Annexure XI(E) (Contd.....)

Sl. No.	Division	Palash	Gaba	Char	Benacher	Karanja	Gillo	Shikakai	Mahua
17.	Rayagada	2.17	—	—	—	182.30	—	—	—
18.	Kalahandi	516.60	12.54	—	170.59	255.20	—	—	—
19.	Phulbani	62.00	5.77	0.04	0.15	—	11.85	0.13	0.21
20.	Sundergarh	159.00	—	—	—	328.97	17.46	—	—
21.	Sambalpur	12433.83	—	—	—	2900.00	0.80	—	—
22.	Nowrangpur	1578.90	36.88	—	—	60.01	43.55	—	0.17
23.	Boudh	309.43	—	—	23.73	—	0.86	—	—
24.	Bonai	25.37	—	—	—	13.50	100.54	—	—
25.	Puri	—	—	—	—	—	—	—	—
26.	Rairakhol	45.15	—	—	—	—	—	—	—
27.	Parlakhemundi	—	—	—	—	—	—	—	—
	Grand Total	**16445.37**	**65.15**	**0.19**	**1375.54**	**4673.56**	**337.56**	**0.63**	**0.38**

Annexure XI (F)

Quantity of EFP sold by Utkal Forest Products Ltd. from 1992-3 to 1994-5

Name of item	*1992-3*	*1993-4*	*1994-5*
Oil			
Gaba	—	—	24.25
Marda	—	—	16.50
Neem	-	533.05	499.00
Banatulsi	132.25	667.15	—
Kamalagundi	—	—	—
Mahua	4036.65	15035.00	5369.15
Polang	—	—	9.55
Karanja	180.10	120.28	1659.10
Mango kernel	644.30	7.00	—
Banakulthi	188.85	—	192.25
Kusum	1282.65	2077.44	2247.90
Acid oil	455.25	1113.45	161.45
Gillo	—	18.54	3.50
Palash	—	—	36.90
Chakunda	—	28.50	—
D.O.C.			
Banatulsi	24839.26	14185.87	50.00
Mahua	29052.39	21569.35	9090.30
Karanja	298.05	3675.90	113.40
Mango kernel	6019.55	1271.10	—
Gaba	198.90	64.60	—
Neem	3175.40	4447.15	619.80
Chakunda	260.05	4448.25	—
Banakulthi	1706.55	—	280.00

Name of item	*1992-3*	*1993-4*	*1994-5*
Kusum	448.85	3494.65	1300.00
Babool	953.70	2103.60	5731.00
Makhana	0.11	—	—
Palash	—	—	120.00
Marda	—	—	100.00
Chardana	—		154.26
Sales			
Seeds			
Sikakai seeds	225.08	95.20	—
Karanja seeds	—	19.00	3440.00
Babool seeds	—	—	—
Banatulsi seeds	557.85	773.85	35700.00
Chakunda seeds	—	—	50100.00
Chiranji seeds	01.11	—	—
Harida seed	1313.58	—	—
Gillo seed	613.45	1153.65	48.75
Benacher seed	11.65	—	—
Indrajaba	20.82	4.00	—
Baghanakha seed	67.85	—	—
Bahada	1955.94	—	—
Anato	155.50	—	69.25
Amla	269.22	—	—
Nageswar	1.10	—	—
Neem	—	120.00	2400.00
Kusum	—	41.00	2680.00
Chardana	—	10.72	1729.48
Mango kernel	—	—	48710.00
Banakulthi	—	—	120.00
Gaba	—	—	100.00
Mahua	—	—	5460.00
Semi-Agricultural Products			
Turmeric			2000.00
Maize			39700.00
Redgram			1000.00

Annexure XI (G)

Government's letters granting lease to UFP Limited

Government of Orissa

Forest, Fisheries and A.H. Department

No. 5F-CS-7/90.29672/FFAH, Dated Bhubaneswar the 4 Dec. 1990

From

Shri G.A. Obaid,

Under Secretary to Government

To

The Pr. Conservator of Forests, Orissa

Sub : Government Policy on lease of Minor Forest Produce items in the State

Ref : (i) This Deptt. Memo No. 9869/FFAH, Dt. 4.3.90

(ii) This Deptt. letter no.23632/FFAH, Dt. 26.9.90

(iii) This Deptt. letter no. 26145/FFAH, Dt. 5.11.90

Sir,

I am directed to say in order to eliminate middlemen from collection and trade of MFP items. Government have decided that all leases of MFP items should be settled with TDCC, Co-operative Societies. Government owned Corporations and joint sector companies. There shall be no private lease in these items. Small- and medium-scale industries depending on MFP items for their raw materials should have tie-up arrangement with the above agencies for supply of raw materials. The collecting agencies namely Orissa Forest Development Corporation/TDCC/AMCS, Tikabali and similar other co-operatives should consider setting up smaller scale industries in the backward areas based on such MFP items in order to provide local employment and economic upliftment of the tribals. In pursuance of the above policy decision of Government the following guidelines may be adopted with regard to settlement of lease of MFP items in the state.

(i) TDCC Ltd. will have the exclusive right to operative lease of 4 MFP items, namely Tamarind, Hill Broom, Honey and Mahua Flower in all the 27 divisions of the State (This Deptt. G.O. No. 28403/FFAH, dated 6.12.89).

(ii) M/s. Utkal Forest Products Ltd., a joint sector Company under IPICOL will operate lease of 29 identified MFP items in all the forest divisions of the State (This Deptt. G.O. No. 23632 dt. 26.9.90 read with G.O. No. 26145 dt. 5.11.90).

(iii) TDCC will operate lease of all the MFP items excluding the 29 MFP items identified for M/s. Utkal Forest Products Ltd. in 19 Forest Divisions namely, Khariar, Nawarangpur, Jeypore, Kalahandi, Rayagada, Nayagarh, Ghumsur North, Ghumsur South, Parlakhemundi, Baripada, Karanjia, Keonjhar, Sundergarh, Deogarh, Bamara, Bonai, Bolangir, Sambalpur and Rairakhol.

(iv) In the remaining 5 divisions, namely, Puri, Dhenkanal, Angul, Athagarh and Athamallik, OFDC Ltd. will operate lease of MFP except the 29 items identified for M/s. Utkal Forest Products Ltd. and 4 items reserved for TDCC Ltd.

(v) AMCS, Tikabali will operate lease of MFP items except the 29 MFP items identified for M/s. Utkal Forest Products Ltd. and 4 items identified for TDCC Ltd. in Phulbani, Baliguda and Boudh forest divisions.

You are therefore, requested to take immediate follow-up action in the matter.

Yours faithfully

Under Secretary to Government

Memo No. 29073/FFAH, dated the 4.12.90

Copy forwarded to all divisional forest officers (Terr.)/All conservators of Forests (Terr.) Managing Director, TDCC/Managing Director, OFDC Ltd./Registrar, Cooperative Societies/Secretary, AMCs, Tikabali/Managing Director, IPICOL for information and necessary action.

Under Secretary to Government

Memo No. 29074/FFAH., dated the 4.12.90

Copy forwarded to Harijan and Tribal Welfare Deptt./Industries Department/Agriculture and Co-operative Deptt. for information and necessary action.

Under Secretary to Government

Memo No. 29075/FFAH, dated the 4.12.90

50 copies to Guard File.

Under Secretary to Government

Annexure XI (H)

Government's letters granting lease to UFP Limited

Copy of Letter No. 5F-M-58/89.23632/FFAH, dated Bhubaneswar the 26.9.1990 from Shri D.S. Mishra, Deputy Secretary to Government of Orissa F.F. & A.H. Department addressed to the Principal Chief Conservator of Forests, Orissa, Bhubaneswar.

Sub : Lease of MFPs to the Utkal Forest Products Ltd., a Joint Sector Company under IPICOL – Lease for the year 1990-91 and 1991-92.

In supervision of this Department letter No. 10046/FFAHY., dated 5.5.90 cancelling the order of lease of MFP items in favour of M/s. Utkal Forest Products Ltd. and in partial modification of this Department G.O. No. 28403, dt. 5.12.89 to the Principal Chief Conservator of Forests, Orissa. I am directed to say that Government have been pleased to grant lease of the following 29 MFP items in favour of M/s. Utkal Forest Products Ltd., a Joint Sector Company set-up in collaboration with IPICOL, for a period of 2 (two) years, i.e., from 1.10.1990 to 30.9.1992, on experimental basis in the first instance.

1. Kusum Seed
2. Neem Seed,
3. Sikkakai Seed
4. Karanja
5. Mahua seed
6. Mango kernel
7. Gabo
8. Palas seed
9. Madra seed
10. Nageswar seed
11. Sheali seed
12. Indrajaba
13. Gilo seed and cover
14. Benacher
15. Banhaldi
16. Bankulthi

17.	Chakunda	18.	Bantulsi
19.	Lado Baguli	20.	Baidanka
21.	Kamalagundi	22.	Baghanakha
23.	Babul seed	24.	Chiranji dana
25.	Ananta,	26.	Basil
27.	Jute seed	28.	Polang
29.	Makahana seed		

2. It may be reiterated that the minimum royalty payable by the Company in respect of MFP items was based on royalty figures of the year 1987-88. Hence, the latest minimum royalty figures relating to these MFP items should be collected and minimum royalty, whichever is higher out of these two figures, should be fixed for the company vide para 8.1 of this department's G.O. No. 28403, dated 6.12.89.

3. The company will have to procures the leased MFP items in its own processing units within two years.

4. All other terms and conditions of the lease shall remain the same as in the G.O. No. 28403, dated 6.12.89.

5. The Joint Sector Company will have to execute an agreement with the State Government prior to operation of the lease.

6. Hence, you are requested to send the draft agreement incorporating the terms and conditions stated above in the earlier G.O. under reference to this Department immediately.

Annexure XII

Procurement of SAP & EFP by TRIFED, Bhubaneswar during 1988-89 to 1994-95
(Qty in MT. value in lakhs)

Sl. No.	Commodity	1988-9		1989-00		1990-1		1991-2		1992-3		1993-4		1994-5	
		Qty.	Value	Qty.	Value	Qty.	Value	Qty.	Value	Qty.	Value	Qty.	Value	Qty.	Value
1.	Niger Seed	1115	81	1297	98	3265	423	500	60	4027	411	71	6	178	18.22
2.	Maize	—	—	—	—	147	30	—	—	—	—	—	—	4730	221.5
3.	Jowar	—	—	—	—	923	18	—	—	—	—	—	—	—	—
4.	Ragi	—	—	—	—	246	2.8	—	—	—	—	—	—	—	—
5.	Blackgram	—	—	—	—	86.3	5.74	—	—	—	—	—	—	—	—
6.	Rajma	—	—	—	—	29	3	—	—	—	—	—	—	—	—
7.	Cashewnuts	—	—	—	—	63.78	8.3	—	—	57	7.4	—	—	82	30.9
8.	Catting	—	—	—	—	46.3	1.7	—	—	—	—	—	—	—	—
9.	Tamarind	—	—	4111	127	—	—	—	—	948	21	—	—	5646	220
10.	Tamarind concentrate	—	—	—	—	0.75	10.197	—	—	—	—	—	—	—	—
11.	Tamarind powder	—	—	—	—	0.3	0.112	—	—	—	—	—	—	—	—
12.	Myrobalans	693	1.44	—	—	300	8.1	—	—	—	—	—	—	519	10.37

13. Mahua seed	—	—	—	—	241	16.87	—	—	—	—	—	—	546	49.14
14. Sal seed	1032	—	—	—	—	—	—	—	—	—	—	—	1447	33.28
15. Sal oil	129	—	—	—	—	—	—	—	—	—	—	—	—	—
16. Sal doc	854	—	—	—	—	—	—	—	—	—	—	—	—	—
17. Gum karaya	—	—	10.57	5.4	47.59	17.96	—	—	—	—	—	—	—	—
18. Kendu leaves	—	—	—	—	84	33.88	—	—	—	—	—	—	—	—
19. Cocoons	—	—	—	—	3208	4.9	—	—	—	—	—	—	—	—
20. Tasar fabric	—	—	—	—	77	0.06	541*	3.68	7.7 mt	0.089	—	—	—	—
21. Handicrafts	—	—	—	—	76	0.043	—	—	—	—	—	—	—	—
22. Chakunda (puwa seed)	—	—	—	—	—	—	—	—	—	—	—	—	5007	158
23. Turmeric	—	—	—	—	—	—	—	—	—	—	—	—	2390	298.86
24. Redgram	—	—	—	—	—	—	—	—	—	—	—	—	100	12
25. Forest caster	—	—	—	—	—	—	—	—	—	—	—	—	10	0.7
26. Karanjia	—	—	—	—	—	—	—	—	—	—	—	—	334	15
27. Kusum	—	—	—	—	—	—	—	—	—	—	—	—	618	30.9
28. Neem	—	—	—	—	—	—	—	—	—	—	—	—	240	10.8
29. Mango kerne	—	—	—	—	—	—	—	—	—	—	—	—	4871	146.13
30. Chiranjee	—	—	—	—	—	—	—	—	—	—	—	—	173	147
31. Banatulsi	—	—	—	—	—	—	—	—	—	—	—	—	3570	214.2
32. Nux vomica	—	—	—	—	—	—	—	—	—	—	—	—	45.7	4.11
33. Cleaning nut	—	—	—	—	—	—	—	—	—	—	—	—	14.74	1
34. Red chillies	—	—	—	—	—	—	—	—	—	—	—	—	766	223.76

Annexure XIII

Volume of 65 items of EFP collected from 1985-6 to 1991-2*

Sl. No.	Name of Produce	Unit	1985-6 Qty.	1986-7 Qty.	1987-8 Qty.	1988-9 Qty.	1989-90 Qty.	1989-90 Gross value (in Rs.)	1990-1 Qty.	1990-1 Gross value (in Rs.)	1991-2 Qty.	1991-2 Gross value (in Rs.)
1.	Bamboo	MT	274528	2424442	NA	260955	265650	49649161	215965	64499392	236940	70040629
2.	Canes	Qtl.	27	69	99	256	3024000 nos.	1824400	896000 nos.	627200	146591 nos	102620
3.	Tamarind	Qtl.	123932	113327	84012	149863	52310	23984498	165213	76590848	94518	42978098
4.	Mahua flower	Qtl.	646514	685971	700936	394575	607375	182212500	449501	134850300	632478	—
5.	Patal garuda	Qtl.	18	19	—	—	—	—	—	—	10	6710
6.	Arrowroot	Qtl.	—	38	7	2	80	176000	13	28600	34	85000
7.	Honey and Wax	Qtl.	1829	94	—	69	21	21000	21	31500	96	144000
8.	Nux vomica	Qtl.	591	1689	1211	173	225	90592	1300	359342	132	67342
9.	Shikakai	Qtl.	120	148	2212	377	274	86985	221	70165	211	66992
10.	Chireita	Qtl.	160	247	118	624	489	48900	10194	1019400	1920	192000
11.	Dhataki flower	Qtl.	2500	2132	178	1645	1415	311300	635	139700	7	1540
12.	Gilo	Qtl.	—	—	—	160	40	6600	138	22780	1330	224250
13.	Banatulsi	Qtl.	—	200	1129	4094	22116	4538560	7347	2175520	6755	2080800
14.	Indrajaba	Qtl.	15	15	90	223	404	126028	224	69868	23	7170
15.	Sunari bark	Qtl.	1473	624	154	—	10	2658	—	—	367	29175
16.	Dhantri bark	Qtl.	350	1970	320	—	50	15334	1350	1166	1650	31350
17.	Antiaa bark	Qtl.	514	118	158	50	—	—	—	—	—	—
18.	Siali fibres	Qtl.	1091	1129	2024	836	3696	1846500	407	203500	1669	834500
19.	K.Creeper	Qtl.	18	39	30	85	32	2400	65	4875	—	—

20. Kathalai	Qtl.	489	—	—	—	—	—	—	—	—	—
21. Mat reed	Qtl.	28858	317	3300	389	—	—	—	—	93440 panel	37376
22. Bena cher	Qtl.	—	—	—	931	1755	366805	—	—	427	89249
23. Kantabandhuni	Qtl.	—	—	—	50	—	—	—	1074	85910	—
24. Sabai grass	Qtl.	2989	4828	3232	2652	70	10500	2712	406800	1740	261000
25. Genduli gum	Qtl.	3756	1516	2805	3019	5979	14947500	1394	348500	2405	601500
26. Other gum	Qtl.	1935	1143	—	512	1160	—	—	—	—	—
27. Sal resin	Qtl.	31	47	—	-213	—	—	—	—	—	—
28. Lac	Qtl.	2704	102	126	149	209	146300	231	161700	2	1400
29. Myrobalans	Qtl.	34014	10758	8356	45630	48150	6075397	53868	6790147	37516	4746147
30. Tasar cocoons	Qtl.	157	812	180	319	12543288 nos	413044	8184931nos	472536	8642809 nos	480079
31. Cleaning/ marking nuts	Qtl.	5070	4808	3774	10502	1178	33982	4363	75851	770	—
32. Mahua seeds	Qtl.	32960	42139	33938	27527	36538	23353560	7134	5123080	11314	7229570
33. Kusum seeds	Qtl.	9051	2637	5084	4943	1159	481390	4529	1189090	12657	2895970
34. Sal seeds	Qtl.	310283	669155	324132	537865	521624	162567600	248988	87146200	398974	139640100
35. Karanja seeds	Qtl.	2729	9491	2057	12707	2239	570190	5499	1254790	379	179590
36. Neem seed	Qtl.	866	2775	2457	802	14148	2466236	3857	850549	2861	694177
37. Cashew nut seed	Qtl.	3562	3316	4119	4184	—	—	1082	2017309	1039	1937381
38. Char seed	Qtl.	366	2422	138	1855	3489	1803815	3230	1669910	893	461675
39. Mango kernel	Qtl.	2025	6788	21399	18842	895	148600	15909	2640900	5700	946200
40. Kendu leaves	Qtl.	(lakh) 3.84	4.28	3.38	3.96	4.45	824443791	3.96	768500000	5.05	515732000
41. Siali leaves	Qtl.	85829	249478	78984	136249	76219	7646152	65691	6593352	48025	4826752
42. Sal leaves	Qtl.	26551	—	14984	28633	7477	57813	1762	4000	—	234901
43. Versinga leaves	Qtl.	3920	12000	—	5080	92	3800	3080	126280	484	16564
44. Khajuri pati	Nos.	—	9460	4016	1974	3000	60000	6830	136600	2800	56000

Annexure XIII (contd...)

Sl. No.	Name of Produce	Unit	1985-6 Qty.	1986-7 Qty.	1987-8 Qty.	1988-89 Qty.	1989-90 Qty.	1989-90 Gross value (in Rs.)	1990-1 Qty.	1990-1 Gross value (in Rs.)	1991-2 Qty.	1991-2 Gross value (in Rs.)
45.	Thorn broom	Qtl.	—	—	—	2574	—	—	—	—	—	—
46.	Broom stick	Nos.	1286550	5388300	1467700	3373000	2284220	5139495	776102	1746229	1556500	3502125
47.	Anata	—	—	—	—	—	—	—	—	—	145	46037
48.	Chakunda seeds	—	—	—	—	1123	123521	187	20579	14697	1616679	—
49.	Bhaghanakhi	—	—	—	—	—	3	171	5	310	97	5779
50.	Babul seeds	—	—	—	—	—	—	—	—	—	236	14412
51.	Fenfena	—	—	—	—	—	—	—	—	—	10	20737
52.	Palasa	—	—	—	—	—	—	—	2	420	25	5630
53.	Gaba	—	—	—	—	—	73	15750	483	104310	28	6060
54.	Banakulthi	—	—	—	—	—	—	—	46	330	1753	140146
55.	Banahaldi	—	—	—	—	—	—	—	50	8100	74	11968
56.	Kamalagundi	—	—	—	—	—	—	—	—	—	77	16970
57.	Marda seeds	—	—	—	—	—	—	—	—	—	89	37655
58.	Baidhanka	—	—	—	—	—	—	—	8	1920	5	1200
59.	Nageswar seeds	—	—	—	—	—	—	—	—	—	1	275
60.	Polanga	—	—	—	—	—	25	16350	—	—	6	4000
61.	Makan seed	—	—	—	—	—	—	—	—	—	56	17280
62.	Palm leaves	—	—	—	—	—	1010	3700	—	—	—	—
63.	Simuli cotton	—	—	—	—	—	1	1624	7	10500	5	9590
64.	Mushroom	—	—	—	—	—	—	—	1007	60420	78	4653
65.	Sisal fibre	—	—	—	—	—	—	—	—	—	230	9156

Annexure XIV

Properties and processing of Soapnuts

Introduction

THE commonly known soapnut trees, Sapindus emarginatus and Sapindus mukorossi are widely distributed in several States like Karnataka, Andhra Pradesh, Orissa, Himachal Pradesh and other northern parts of India. Soapnuts are well known for their detergent property which is due to the presence of triterpene saponins. The Saponoin content of the soapnuts varies from 10-18 per cent.

Uses and market demand

The saponins have various uses in detergents shampoos, textiles, foods, pharmaceuticals and photofilm industry. The saponin requirement of the photofilm industry is 6 tons per annum, presently being totally imported.

A process is developed for isolation and purification of saponins from the soapnuts. The soapnuts are cut and separated into pericarp and seeds. The pericarp is used in the extraction of saponin. The seeds contain kernel which has about 36 per centof oil which may be utilised in the soap industry.

Special features of the process

The process of isolation from soapnuts involves the following operations:

1. Extraction of the pericarp of soapnuts with a solvent by counter current method.
2. Precipitation of the saponin from the extract.
3. Purification of saponin.

4. Drying to get solid saponin.

Solvents are recovered for reuse in the process. All the raw materials and chemicals required are available indigenously.

Characteristics of the saponin

Sapogenins (around)	:	21%
Ash	:	1%
Volatiles	:	5%
Surface tension	:	39-42 dynes/cm
Physical appearance	:	Light yellow/brown powder
Emulsifying Property (with rape seed oil)	:	0.07 - 0.18
Ph. (1% aq. solution)	:	5-6

The process is developed on 4 kg batch scale of the peicarp. The recovery of the saponins 85-90 per cent of the amount present in the pericarp.

Equipment (based on 100 kg per day pericarp processing)

1.	Extractors (MS)	500 lit.	4 Nos.
2.	Precipitation tanks	400 lit.	2 Nos.
3.	Solvent extractors jacketed with stirrer	100 lit.	1 Nos.
4.	Drying tanks	100 lit.	2 Nos.
5.	Distillation unit	100 lit	1 Nos.
6.	Storage tank	500 lit	4 Nos.
7.	Boiler	—	1 No
8.	Spray dryer	—	1 No.
9.	Vacuum pump	—	1 No.

Utilities

1. Water, 2. Stream, 3. Electricity

Economics (based on the price of 1989-90 is in Rs. Lakh)

Fixed capital : Rs. 15.12

Working capital	:	Rs. 3.46 (3.01)
Capital investment	:	Rs. 18.58 (18.13)
Production cost of saponin	:	Rs. 14.36 (12.66)
Profitability after taxation	:	25.95% (14.74%)

Figures in brackets refer to S. Emarginatus.

***Source** : Regional Research Laboratory, Bhubaneswar.

Annexure XV

NGOs seek changes in Forest Bill for its bureaucratic overtones

More than 1,500 non-governmental organisations (NGOs) all over the country have called for drastic changes in the draft Forest Bill prepared by the Ministry of Environment and Forests to replace the Indian Forest Act, 1927. The request is with a view to involve people at the grass-roots in the protection and preservation of forests.

The amended draft Forest Bill prepared by the National Campaign Committee for People's Resources, was presented to the Minister for Environment and Forests in the last week of August by committee chairman Walter Fernades of the India Social Institute.

The NGOs found that the draft Bill circulated by the Ministry not only retained all the objectionable and archaic sections of the present Act, but also proposed to add many more sections and clauses, thus giving greater power to the forest bureaucracy and negating the rights of the people.

They further felt that the draft did not follow the resolution on the National Forest Policy, 1988, in letter and spirit. The National Forest Policy resolution called for 'meeting the basic needs of the people, especially, fuelwood, fodder and small timber for the rural and tribal people and maintaining the intrinsic relationship between forests and the tribals and other poor people living in and around the forests by protecting their customary rights and concessions on the forests.'

It was also felt that the draft will not be able to achieve the main objective of the National Forest Policy of 'creating a massive people's movement with the involvement of women for achieving these objectives and to minimise pressure on existing forests.'

The official draft, the NGO organisations found, also did not take into consideration of the important policy initiative of Joint Forest Management (JFM) as reflected in the circular issued by the Ministry of Environment and Forests in June, 1990.

The Bill, on the other hand, has provisions that will act as disincentives to form village forests as a modified form of JFM, for, it empowers the State Governments to take back the management of village forests from the local body.

The NGOs submitting the amended bill reminded that the nation is facing an acute crisis on the forest front. The tree cover is fast declining, creating numerous ecological problems in addition to creating hardships to adivasis and other communities dwelling in forests. 'What is needed is a very radical and innovative forest legislation that can generate a mass movement for forest development and preservation', they added.

The Department of Environment and Forests alone cannot generate such a people's movement, especially if it wants to have a monopoly in decision-making and monitoring, the NGOs felt.

The Department of Forests will have to accept a new role of a friend, philosopher and guide a friend, philosopher and guide with its technical expertise and not that of a master who takes all the decisions.

People's movements, the NGOs stressed, are generated by the people and never by a government department. The last statement in the national policy resolution that 'forests are a national asset to be protected and enhanced for the well-being of the people and the nation' must be fully accepted, the NGOs felt.

The NGOs found that the Bill did not take into account the revolutionary spirit of the 73rd and 74th Amendments to the Constitution of India that promises to usher in a new era of decentralisation of power in India. The main purpose of the amendments suggested is to emphasise on the supremacy of people in governance.

The NGOs suggested that the spirit must form the basis for the management of forests. Only people can preserve and develop their natural resources and they must be encouraged to do so by assigning rights and duties, they added.

The official Bill, on the other hand, centralises decision-making in all matters relating to forests. The Central Government is taking over even the powers of the various States to make rules according to local needs and circumstances. Greater centralisation will mean that local organisations and movements will no longer be able to pressurise or bargain with the State Governments, the NGOs pointed out.

In respect of reserved and protected forests, the NGOs felt that the procedures have been made even more complicated and arbitrary and no mention has been made of the rights of the people who settled in the forests long ago.

Annexure XVI

Guidelines on essential forest produce

Collection/Procurement

1. Types of Essential Forest Produce (EFP) available in the Division. (Tick Mark the items given in the list enclosed)
2. Important areas in the Division where EFP (Tick marked) are found. (Show in a map the distribution of different items).
3. Quantity (in quintals), produced/collected (Mention in the list latest information).
4. Communities collecting EFP.
5. Collection Agency
   (a) Self-collection
   (b) Collection through any agency (businessman, middleman, OFDC, TDCC, LAMPS, any others).
6. Methods of collection of different EFP.
7. Any arrangement for training on modern technique of collection.
8. Collection season.
9. Monopoly of any agency for procurement of any EFP and advantages and disadvantages of monopoly procurement.
10. Is there any practice of replenishment of plants observed among those who gather EFP ?
11. What items of EFP collected individually and what collectively and why it is so ?
12. Are there any unknown items of EFP which are left out and not collected now ?

13. If so, what are those items and what is their utility ?
14. What techniques are used for collection of gum Karaya ? (Is death tapping practised ?)
15. What systems are followed for procurement of EFP ?
16. What difficulties are faced in collection and how are they sorted out ?
17. Are there any estimates made of quantities of any EFP available for procurement ?
18. Are any targets fixed for collection in terms of quantity ?
19. Is there any such practice followed in respect of procurement of specific items by specific agencies. (say for example, procurement of turmeric by TDCC) ?
20. Is there any item of EFP nationalised ? If so, is it more profitable than other system ?
21. Is any mapping of forest produce done ? (If so, furnish a copy of the map)
22. Does inter-State co-ordination and uniformity in the forest policy help better procurement ?
23. Are the rights of tribals codified in respect of collection of EFP for domestic consumption ?
24. What rules and regulations are observed in respect of procurement by the government ?

Storage

1. Where are storage and transportation facilities available ? Indicate these facilities in a map including controlling agencies.
2. What arrangements are available for storage of EFP and which agencies have provided such facilities ?

Processing

1. Is there any preliminary processing of any EFP done ?
2. If so, what types of EFP are processed and what methods and mechanisms of processing used ?
3. Which agency is involved in processing ?

4. Are the EFPs used locally for any purpose ?
5. Are any expertise and skill needed for'

   (a) Collection, (b) Preservation, (c) Storage, (d) Processing, (e) Marketing, and (f) Manufacturing, of EFP.
6. Is any training needed for processing, manufacturing and marketing, etc. ?
7. Is there any costing done of collection, processing, manufacturing, transportation of products and what are the credit needs ?
8. How to assess economic viability of different items of EFP and has it been done for any item ?
9. What role do the NCDC, NAFED play in procurement, processing, storing and marketing ?

Pricing

1. Price of EFP

   (a) Price paid by local businessmen to the persons collecting EFP.

   (b) Price paid by OFDC/TDCC/LAMPS to persons collecting EFP.

   (c) Normal market price at weekly markets.
2. How is remunerative price paid to the gatherers determined ?
3. How is sale price of the EFP determined ?
4. Is there any difference noticed in price offered by the private traders and institutionalised or co-operative agencies ?
5. If so, what possible reasons and what are the consequences of such price differences ?
6. Is there any provision of support price for any MFS and what are possible reasons for such provisions and how does it help in full scale procurement of such EFP ?
7. Does the Forest Department charge any royalty on sale price for maintenance of any NFP species ?

8. Is the system of differential pricing followed for different grades of a EFP. (Say for example gum or arrowroot.)

Marketing

1. Name/location of weekly markets existing in the division.
2. Types of EFP sold at different markets in different seasons.
3. Any co-operative organisation existing for business in EFP.

Organisations

1. What rights do the tribals enjoy in respect of collection of EFP ?
2. What measures are taken to ensure remunerative price to tribals ?
3. What institutional arrangements are in existence for marketing of EFP and for tackling different matters, connected with EFP collection, grading, collection of marketing intelligence, export and improving the resource base ?
4. What are the various handicaps which stand in the way of realising remunerative price for different items of EFP collected by the tribals, (lack of marketing intelligence, lack of storage and transportation facilities, differential rates of procurement, etc.) ?
5. What roles do the LAMPS, TRIFED, NAFED play in resolving difficulties, arising in respect of collection, grading, storage, marketing, transportation, and export of EFP and offering remunerative price to the tribals ?
6. What assistance do these institutional organisations offer to save the tribals from exploitation by private traders ?
7. What support and assistance do these organisations receive from both Central and State Governments ?
8. Are there any steps taken to bring collection of EFP in the nationalised sector and to abolish the royalty being levied on the EFP ?
9. Are there any steps taken by the TRIFED for training the tribals in scientific collection of EFP and also for strengthening

LAMPS to make it a viable collection agency ?

10. Is the capital base of the TRIFED strong enough to venture marketing of large volume of different items of EFP ?
11. To what extent has the National Forest Policy, 1988 which profess full protection of the rights and concessions to the tribals been put into practice ?
12. What sort of organisation is needed to be set-up at the grass-root level to ensure remunerative price to the tribals and prevent exploitation ?
13. How is LAMPS working in the matter of helping procurement of EFP ? What changes or reforms are needed in the LAMPS to make it efficient and effective ?
14. How is TDCC working in the matter of procurement and business in EFP ? What organisational; reforms are needed to make it more effective ?

EFP based industries

Name and location of EFP based industries in the area:

(a) Name

(b) Location

(c) EFP used and quantity used

(d) Product manufactured

(e) Marketing and produce

(f) Difficulties

(g) Financial input, agency providing funds, credit system.

(h) Participation and employment of tribals in industries.

Annexure XVII

Utilisation of Essential Forest Produce

Fibres and Flosses

TEXTILE FIBRES

Uses of textile fibres are as follows :

Products like fabrics, netting and cordage are important. Cotton, flax, jute and hemps are some of the important commercial fibres.

PLATING/WEAVING FIBRES

The products which are manufactured by using these hard fibres are straw mats, chair seats, baskets and other wicker works.

BRUSH FIBRES

These are used for making brushes and brooms.

FILLING FIBRES

The uses to which these hard material are put, include upholstery, mattresses, barrels, and wall boards.

PAPER-MAKING FIBRES

Wood fibres are used for making pulp and paper.

NATURAL FABRICS

Barks extracted from trees are softened and used for making cloths which were used in ancient time by primitive tribes.

Grasses, reeds and canes

GRASSES

Grasses are used as fodder, for making textile fibres, cordage, matting, paper and pulp. Some other uses of grass are *khus-khus* making, brooms and for making seats, chairs, baskets and screens.

CANES

Some of the uses of canes are as follows:

Suspension bridges are made by using canes. Frames for furniture, walking sticks, umbrella handles are also made out of canes. Backs of sofa and chairs are made of canes. The canes are also used in wicker works. Sports goods are manufactured with the use of canes.

Essential oils

GRASS OILS

Citronella oil

It is a mosquito repellent is made by mixing it with cedar oil. It is also used in soaps, floor waxes, insecticides and perfumes.

Ginger grass oil

A number of chemicals are obtained from this oil. It is also used in making scents, soaps because of its sweet smell.

Lemon grass oil

This oil has a lemon odour, and is used in manufacturing perfumes and soaps. The essential ingredients are ionones, methylionones and Vitamin A. Citral is isolated from lemon oil and used for making perfumes. Lemon oil is used to prepare pharmaceuticals such as medicines for relieving pains and mosquito repellents.

Palmarosa oil

Palmarosa oil is needed as base for perfumes and cosmetics. It is used in tobacco for its smell. From this oil a mosquito repellent is made. It has a geranium odour and is a pale yellowish liquid. The Palmarosa oil is a constituent part of many medicines and it mixes well with other soap perfumes.

Khus oil

The Khus oil is used in the manufacture of perfumes, cosmetics and soaps. Flavouring of beverages is done with the use of this oil. It mixes well with sandalwood and rose oil.

WOOD OILS

Agar oil

Its smell is like that of sandalwood oil. Agar is popular for burning as incense and *agarbati* (incense stick). This oil is used to dilute sandalwood oil to produce agar perfume. It is of great commercial value and used for making perfumes.

Sandalwood oil

Sandalwood oil is well known as a scented oil. It mixes well with rose, clove, lavender and other oils. It has a pleasant smell.

LEAF OILS

Camphor oil

Camphor oil has a medicinal value. It is used to cure cough and cold. It is used in the manufacture of liniments, perfumes and insecticides.

Eucalyptus oil

Its characteristic is its aromatic odour and peculiar spicy taste. It is of medicinal value since it is used to cure cold. It is used to manufacture low cost perfumes.

ROOT OILS

Costus oil

This oil has an extremely tenacious smell. This oil is well-known for its medical properties. It is used to make many types of perfumes and scents.

Keora oil

It is used for the preparation of many perfumes such as *Attar* and

Keora water. It is also used in the preparation of many cosmetics and toilet items.

FIXED OILS AND WAXES

Fixed oils are used for many purposes, such as, for cooking, as a butter substitute for margarines and salad dressings. They are also used for the manufacture of soaps, cosmetics and toilet preparations and for paints, varnishes and lubricants.

Madhuca indica Gmel.

Oil is extracted from the seeds of *Madhuca indica*. It is used for the manufacture of laundry soaps. It is edible and used in cooking and for burning lamps it is used locally. It serves as a hair tonic and has medicinal value. The oil cake is neither used as a cattle feed nor as a fertiliser, since it contains poisonous saponin.

Madhuca longifolia (Koenig) Macbr.

Oil is extracted from the seeds. It is used for manufacturing soaps. It is used for cooking and lighting and has medicinal properties. Its seed cake is used as a manure, but not as cattle fodder.

Schleichera oleosa (Lour.) Oken

The seeds yield oil which has many uses.

It is edible and used for cooking. This oil is also used for lighting. It is used for cleaning and for the growth of hair.

Shorea robusta Gaertn.f.

Oil is extracted from the kernels of the seeds of the plant. It is used for cooking food, for illumination, and for the manufacture of soaps, in which case Sal butter is used.

Azadirachta indica A. Juss.

The seeds of this tree yield oil. It has medicinal value. It is used for burning lamps for lighting. It is mixed with other oils and fats for the manufacture of washing soaps.

There are many other trees which yield oil from the seeds. Some of such oil-yielding species are :

1. Arakha (*Calotropis gigantea* R. Br.)
2. Agara (*Argemone mexicana* Linn.)
3. Ashwagandha (*Withania somnifera* (L.) Dunal.)
4. Bhalia (*Semecarpus anacardium* (L.) Linn.)
5. Cashew nut (*Anacardium occidentale* Linn.)
6. Jatropha (*Jatropha curcas* Linn.)
7. Karanja (*Pongamia glabra* Vent.)
8. Mango kernel (*Mangifera indica* Linn.)
9. Neem (*Azadirachta indica* A. Juss.)
10. Oil Palm (*Elaeis guinensis* Jacq.)
11. Polang (*Calophyllum inophyllum* Linn.)
12. Sirisa (*Albizia procera* (Roxb.) Benth.)

The oil of these seeds are used for lighting and some of them are used for medicinal purposes and also for making soaps.

Tans and dyes

TANS

Wood tans

Acacia nilotica (L.) Willd. ex Del.

The bark of this tree yields an important tannin. The bark from older tree is richer in tannin and has a deeper colour. The tannin of this species is most suitable for heavier leather.

Fruit tans

Caesalpinia coriaria (Jacq.) Willd

The tannin is present in the pods of the plant.

The pods of *Caesalpinia coriaria* have a rich tannin content. It produces a leather that is affected by weather conditions. The tannin is blended with other materials to remove its bad effects on leather. The tannin develops red-colouring matter which is bad for tanning.

Terminalia chebula Retz.

The fruits of this plant yield a tannin. This tannin is popular and widely used. Various types of animal skins and hides can be tanned with this tannin. It is an ingredient for the manufacture of ink. Myrobolans are used in the dyeing of cotton and for the weighting of black silk.

DYES

Wood dye

Brazilian dye

It is a dye obtained from the wood of *Caesalpinia sappan* Linn. It is soluble both in water and in alcohol and is used for dyeing cotton, silk and wool fabrics. This dye produces a purple shade in combination with indigo.

Flower and fruit dye

Mallotus phillippensis (Lam.) Muell. -Arg.

The fruits of this plant produce a red dye. Rottlerin is the colouring agent of this dye. It is used for dyeing silk and other fabrics. It is edible since it is used for icecream and drinks. This dye is also used for colouring oil and soaps.

Leaf dyes

Indigofera tinctoria Linn.

Indigo as it is called, is the dye which is extracted from the leaves of this plant. The leaves contain a colourless soluble glucoside which oxidises to form the insoluble indigo. This dye has remarkable permanency and strength. It is used for dyeing fabrics.

Gums and resins

Uses of gums

A high grade variety of gum transparent in colour has a high adhesive power and is used in the clarification of liquor; finishing of silk and in the preparation of water colours. Medium grade gums are used in confectionery, pharmaceuticals, printing-inks, in-sizing and finishing textile fabrics, and dyeing and colouring. Low-grade gums are used as adhesives, and in the paint industry.

Acacia nilotica (L.) Willd. ex Del.

The gum has a property of getting hardened quickly into round tears. The gum is used for the preparation of medicines. It is also used for paper manufacture and in the dyeing and sizing of silk and cotton fabrics.

Anogeissus latifolia (Roxb. ex DC.) Wall. ex Guill & Perr.

This gum is identifiable as its tears are round, opaque externally and transparent internally. They do not crack nor they develop fissures. It is partly soluble in water and forms a colourless mucilage. It is used as gum Arabic substitute. It is used in the pharmaceutical and paper industry. It has some other uses such as it is used as a drilling mud-conditioner in oil exploration and as a water absorbent or desiccant in the explosives industry.

RESIN

Uses of resin

The main uses of resin are the preparation of different types of varnish such as oil varnish and spirit varnish. Resins are used to make soaps and make coatings water-proof. Resins are used in medicine and pharmacy. There are also various other uses of resins.

Shorea robusta Gaertn.f.

Oleo-resin (Lal Dhuma) is produced from this species. When the tree is blazed, resin exudes out. It solidifies quickly on exposure. The months of September to November give maximum resin. A gap of two weeks between the blazes gives more resin. The oleo-resin is opaque, brittle, a light-yellow substance and it is sometimes used in varnishes and paints. Oleo-resin has some medicinal value and is used as a skin ointment.

Drugs and spices

There are many herbal substances which are used for medicinal purposes. There are many publications giving details of medicinal herbs. The bibliography gives name of some publications.

Leaves

Bauhinia racemosa Lam.	: The country bidis (tobacco) are wrapped with the leaves.
Bauhinia vahlii Wt. & Arn.	: The leaves are used for making leaf utensils.
Butea monosperma (Lam.) Taub.	: The leaves are used for making cups and plates.
Corypha umbraculifera Linn.	: The leaves are used in thatching.
Livistona chinensis (Jacq.) R.Br.	: Used for covering palanquins and roofs of boats and thatching houses.
Nipa fruticans Wurmb. ex Bl.	: The leaves are used for thatching and fencing houses.
Tectona grandis Linn.f.	: The leaves of this species are used for packing and thatching houses.

Edible wild plants

The starchy rhizomes, roots and aerial tubers are eaten after boiling. Mango, Jackfruits and many other fruits and Mahua flowers are eaten. Pulp of fruits of perennial plants, particularly woody species and mango kernels are eaten.

Aegle marmelos (Linn.) Corr.	: The fruit of this species is eaten raw or ripe.
Borassus flabellifer Linn.	: It gives juice — a refreshing drink. Tal-misri prepared from sugar gives relief from coughs.
Caryota urens Linn.	: The pith is eaten when food is scarce. It is tapped for salap juice.

Dioscorea belophylla Voigt ex Haines : The tuber is eaten as vegetable.

Diospyros melanoxylon Roxb. : The fruits are eaten.

There are many other forest species giving edible fruits, seeds, juice, tubers and flowers which are eaten.

Appendix

List of Botanical Names

Part I

Botanical name	Oriya name
Abrus precatorius Linn.	Kaincha, Gunja
Acacia catechu (L.f.) Willd.	Khaira
Acacia concinna (Willd.) DC.	Sikakai
Acacia leucophloea (Roxb.) Willd.	Gohira
Acacia modesta Wall.	Phulai (H)
**Acacia nilotica* (L.) Delile	Babul
Acacia sinuata (Lour.) Merr.	Kochi (H)
Acacia senegal Willd.	Kumta (H)
Adina cordifolia (Roxb.) Hook.f.ex Brandis	Halanda, Kuruma
Aegle marmelos (Linn.) Corr.	Bela
Agave cantula Roxb.	Baramasi
Agave sisalana Perr. ex Engelm.	Sisal fibre
Albizia chinensis (Osbeck) Merr.	Kanujera (H)
Albizia lebbeck (L.) Benth.	Sirisa
Albizia procera (Roxb.) Benth.	Dhala Sirisa
Amorphophallus campanulatus (Roxb.) Decne	Olua
Anacardium occidentale Linn.	Kaju
Andropogon ascinodis Clarke	N.A.
Andropogon pumilus Roxb.	N.A.
Anogeissus acuminata (Roxb. ex DC.) Guill & Perr.	Pansi
Anogeissus latifolia (Roxb. ex DC.)	Dhaura

Wall. ex Guill & Perr.	
Andrographis paniculata (Burm. f.) Wall. ex Nees	Kalmegh
Argemone mexicana Linn.	Agara
Artocarpus heterophyllus Lam.	Panas
Artocarpus lakoocha Auct. or *Artocarpus lacucha* Roxb. ex Buch.-Ham.	Jeuta
Azadirachta indica A. Juss.	Neem
Bambusa arundinacea (Retz.) Willd.	Kanta Baunsa
Bassia latifolia Roxb.	Mahua
Bauhinia retusa Roxb.	Kanchana
Bauhinia racemosa Lam.	N.A.
Bauhinia vahlii Wt. & Arn.	Siali
Boswellia serrata Roxb. ex Colebr.	Salai
Bothriochloa pertusa (L.) A. Camus	N.A.
Bombax ceiba Linn.	Simuli
Borassus flabellifer Linn.	Tala
Brachiaria mutica (Forssk.) Stapf	N.A.
**Buchanania lanzan* Spreng	Char seeds
**Butea monosperma* (Lam.) Taub.	Palas
Caesalpinia coriaria (Jacq.) Willd.	N.A.
Caesalpinia decapetala (Roth) Alston	N.A.
**Caesalpinia digyna* Rottl.	Gillo seeds
Caesalpinia sappan Linn.	Bakam (H)
Cajanus cajan (L.) Huth.	Kandula
Calamus guruba Buch.-Ham.	N.A.
Calamus rotang Linn.	Betta
Calamus viminalis Willd.	N.A.
Calophyllum inophyllum Linn.	Polang
Calotropis gigantea R. Br.	Arakha
Canarium strictum Roxb.	N.A.

Canavalia gladiata (Jacq.) DC.	Bada simba
Carissa spinarum Linn.	Khirkoli
Caryota urens Linn.	Salap
Cassia fistula Linn.	Sunari
**Cassia siamea* Lam.	Chakunda
Cassia tora Linn.	Chakunda
Casuarina equisetifolia Linn.	Jhaun
Ceriops decandra (Griff.) Ding Hou	Garani
Chrysopogon aciculatus (Retz.) Trin.	Guguchia
Corypha umbraculifera Linn.	Amada
Curcuma aromatica Salisb.	Ban Haldi
Curcuma angustifolia Roxb.	Palua
Cymbopogon martinii (Roxb.) Wats.	Rosha grass
Cymbopogon citratus (DC.) Stapf	N.A.
Cymbopogon flexuosus (Nees ex Steud) Wats.	N.A.
Cyperus compressus Linn.	N.A.
Dendrocalamus strictus (Roxb.) Nees	Baunsa
Dichanthium annulatum (Forssk.) Stapf	N.A.
Dioscorea belophylla Voigt ex Haines	N.A.
Dioscorea bulbifera Linn.	Pita Alu
Diospyros malabarica (Desr.) Kostel.	Mankada kendu
Diospyros melanoxylon Roxb.	Kendu
Elaeis guineensis Jacq.	N.A.
Emblica officinalis Gaertn.	Amla
Eulaliopsis binata (Retz.) Hubbard	Sabai grass
Ficus benghalensis Linn.	Bara
Ficus glomerata Roxb.	Dimiri
Ficus religiosa Linn.	Aswatha
Garuga pinnata Roxb.	Arinu, Mohi
Gmelina arborea Roxb.	Gambhari
Hemidesmus indicus (Linn.) R.Br.	Ananta

Holarrhena antidysenterica Wall. ex A. DC	Indrajaba
Indigofera tinctoria Linn.	Nila
Jatropha curcas Linn.	Baigaba
Kydia calycina Roxb.	Kapasia
Lagerstroemia parviflora Roxb.	Sidha
Lawsonia inermis Linn.	Mehendi
Livistona chinensis (Jacq.) R.Br.	N.A.
Madhuca indica Gmel.	Mahula
Madhuca longifolia (Koenig) Macbr.	Mahula
Mallotus philippinensis (Lam.) Muell.s-Arg.	Kamala gundi, Rohini flower
Mangifera indica Linn.	Amba
Martynia diandra Glox.	Baghanakhi
Mesua ferrea Linn.	Nageswar flower
Michelia champaca Linn.	Champa
Morinda pubescens Sm.	Achu
**Mucuna monosperma* DC. ex, Wight	Baidanka
Murraya koenigii (L.) Spreng.	Versunga leaves
Nipa fruticans Wurmb. ex Bl.	Hentala
Ocimum basilicum Linn.	Sugandha/ Landa baguli
**Ocimum gratissimum* Linn.	Bana Tulsi
Oroxylum indicum (Linn.) Benth.	Fenfena
Panicum antidotale Retz.	N.A.
Phoenix sylvestris (L.) Roxb.	Khajuri
Polyalthia longifolia (Sonn.) Thw.	Debadaru
Pongamia glabra Vent	Karanja
Pongamia pinnata (L.) Pierre	Karanja
Pterocarpus marsupium Roxb.	Bija
Punica granatum Linn.	Bedana
Rauvolfia serpentina (L.) Benth. ex Kurz	Patal garuda
Saccharum munja Roxb.	Sarakanda

Sapindus emarginata Vahl	Ritha phala
Schleichera oleosa (Lour.) Oken	Kusum
Semecarpus anacardium L.f.	Bhalia
Shorea robusta Gaertn.f.	Sala
Sida cordata (Burm.f.) Brossum	Bajramuli
Spondias pinnata (L.f.) Kurz	Ambada, Salma seeds
Sterculia urens Roxb.	Genduli
Sterculia villosa Roxb. ex DC.	Kodala
Strychnos nux-vomica Linn.	Kochila
Strychnos patatorum L.f.	Nirmala
Symplocos racemosa Roxb.	Lodha
Syzygium cumini (L.) Skeels	Jamun
Tamarindus indica L.	Tentuli
Tectona grandis L. f.	Saguan
Tephrosia purpurea (L.) Pers.	Bana nila
Terminalia arjuna (Roxb. ex DC.) Wt. & Arn.	Arjun
Terminalia chebula Retz.	Harida
Terminalia bellirica (Gaertn.) Roxb.	Bahada
Terminalia tomentosa (Roxb. ex DC.) Wt. & Arn.	Asana
Themeda arundinacea (Roxb.) Ridley	Chana
Thysanolaena maxima (Roxb.) Kuntze	Phool Jhadu
Typha angustata Bory & Chaub.	N.A.
Vanda tessellata (Roxb.) Hook. ex G.Don	Rasna
Vetiveria zizanioides (L.) Nash	Benacher
Withania somnifera (L.) Dunal	Aswagandha
Woodfordia fruticosa (L.) Kurz.	Dhatuki
Ziziphus mauritiana Lam.	Barakoli

Part II

Oriya name	Botanical name
Achu	*Morinda pubescens* Sm.
Agara	*Argemone mexicana* Linn.
Amada	*Corypha umbraculifera* Linn.
Amba	*Mangifera indica* Linn.
Ambada, Salma seeds	*Spondias pinnata* (L.f.) Kurz
Amla	*Emblica officinalis* Gaertn.
Ananta	*Hemidesmus indicus* (Linn.) R.Br.
Arakha	*Calotropis gigantea* R. Br.
Arinu, Mohi	*Garuga pinnata* Roxb.
Arjun	*Terminalia arjuna* (Roxb. ex DC.) Wt. & Arn.
Asana	*Terminalia tomentosa* (Roxb. ex DC.) Wt. & Arn.
Aswagandha	*Withania somnifera* (L.) Dunal
Aswatha	*Ficus religiosa* Linn.
Babul	**Acacia nilotica* (L.) Delile
Bada simba	*Canavalia gladiata* (Jacq.) DC.
Baghanakhi	*Martynia diandra* Glox.
Bahada	*Terminalia bellirica* (Gaertn.) Roxb.
Baidanka	**Mucuna monosperma* DC. ex, Wight
Baigaba	*Jatropha curcas* Linn.
Bajramuli	*Sida cordata* (Burm.f.) Brossum
Bakam (H)	*Caesalpinia sappan* Linn.
Ban Haldi	*Curcuma aromatica* Salisb.
Bana nila	*Tephrosia purpurea* (L.) Pers.
Bana Tulsi	**Ocimum gratissimum* Linn.
Bara	*Ficus benghalensis* Linn.
Barakoli	*Ziziphus mauritiana* Lam.

Baramasi	*Agave cantula* Roxb.
Baunsa	*Dendrocalamus strictus* (Roxb.) Nees
Bedana	*Punica granatum* Linn.
Bela	*Aegle marmelos* (Linn.) Corr.
Benacher	*Vetiveria zizanioides* (L.) Nash
Betta	*Calamus rotang* Linn.
Bhalia	*Semecarpus anacardium* L.f.
Bija	*Pterocarpus marsupium* Roxb.
Chakunda	**Cassia siamea* Lam.
Chakunda	*Cassia tora* Linn.
Champa	*Michelia champaca* Linn.
Chana	*Themeda arundinacea* (Roxb.) Ridley
Char seeds	**Buchanania lanzan* Spreng
Debadaru	*Polyalthia longifolia* (Sonn.) Thw.
Dhala Sirisa	*Albizia procera* (Roxb.) Benth.
Dhatuki	*Woodfordia fruticosa* (L.) Kurz.
Dhaura	*Anogeissus latifolia* (Roxb. ex DC.)
Dimiri	*Ficus glomerata* Roxb.
Fenfena	*Oroxylum indicum* (Linn.) Benth.
Gambhari	*Gmelina arborea* Roxb.
Garani	*Ceriops decandra* (Griff.) Ding Hou
Genduli	*Sterculia urens* Roxb.
Gillo seeds	**Caesalpinia digyna* Rottl.
Gohira	*Acacia leucophloea* (Roxb.) Willd.
Guguchia	*Chrysopogon aciculatus* (Retz.) Trin.
Halanda, Kuruma	*Adina cordifolia* (Roxb.) Hook.f.ex Brandis

Harida	*Terminalia chebula* Retz.
Hentala	*Nipa fruticans* Wurmb. ex Bl.
Indrajaba	*Holarrhena antidysenterica* Wall. ex A. DC
Jamun	*Syzygium cumini* (L.) Skeels
Jeuta	*Artocarpus lakoocha* Auct.
	or *Artocarpus lacucha* Roxb. ex Buch.-Ham.
Jhaun	*Casuarina equisetifolia* Linn.
Kaincha, Gunja	*Abrus precatorius* Linn.
Kaju	*Anacardium occidentale* Linn.
Kalmegh	*Andrographis paniculata* (Burm. f.) Wall. ex Nees
Kamala gundi,	*Mallotus philippinensis* (Lam.)
Kanchana	*Bauhinia retusa* Roxb.
Kandula	*Cajanus cajan* (L.) Huth.
Kanta Baunsa	*Bambusa arundinacea* (Retz.) Willd.
Kanujera (H)	*Albizia chinensis* (Osbeck) Merr.
Kapasia	*Kydia calycina* Roxb.
Karanja	*Pongamia glabra* Vent
Karanja	*Pongamia pinnata* (L.) Pierre
Kendu	*Diospyros melanoxylon* Roxb.
Khaira	*Acacia catechu* (L.f.) Willd.
Khajuri	*Phoenix sylvestris* (L.) Roxb.
Khirkoli	*Carissa spinarum* Linn.
Kochi (H)	*Acacia sinuata* (Lour.) Merr.
Kochila	*Strychnos nux-vomica* Linn.
Kodala	*Sterculia villosa* Roxb. ex DC.
Kumta (H)	*Acacia senegal* Willd.
Kusum	*Schleichera oleosa* (Lour.) Oken
Lodha	*Symplocos racemosa* Roxb.

Mahua	*Bassia latifolia* Roxb.
Mahula	*Madhuca indica* Gmel.
Mahula	*Madhuca longifolia* (Koenig) Macbr.
Mankada kendu	*Diospyros malabarica* (Desr.) Kostel.
Mehendi	*Lawsonia inermis* Linn.
Nageswar flower	*Mesua ferrea* Linn.
Neem	*Azadirachta indica* A. Juss.
Nila	*Indigofera tinctoria* Linn.
Nirmala	*Strychnos patatorum* L.f.
Olua	*Amorphophallus campanulatus* (Roxb.) Decne
Palas	**Butea monosperma* (Lam.) Taub.
Palua	*Curcuma angustifolia* Roxb.
Panas	*Artocarpus heterophyllus* Lam.
Pansi	*Anogeissus acuminata* (Roxb. ex DC.) Guill & Perr.
Patal garuda	*Rauvolfia serpentina* (L.) Benth. ex Kurz
Phool Jhadu	*Thysanolaena maxima* (Roxb.) Kuntze
Phulai (H)	*Acacia modesta* Wall.
Pita Alu	*Dioscorea bulbifera* Linn.
Polang	*Calophyllum inophyllum* Linn.
Rasna	*Vanda tessellata* (Roxb.) Hook. ex G.Don
Ritha phala	*Sapindus emarginata* Vahl
Rohini flower	Muell.s-Arg.
Rosha grass	*Cymbopogon martinii* (Roxb.) Wats.
Sabai grass	*Eulaliopsis binata* (Retz.) Hubbard
Saguan	*Tectona grandis* L. f.

Sala	*Shorea robusta* Gaertn.f.
Salai	*Boswellia serrata* Roxb. ex Colebr.
Salap	*Caryota urens* Linn.
Sarakanda	*Saccharum munja* Roxb.
Siali	*Bauhinia vahlii* Wt. & Arn.
Sidha	*Lagerstroemia parviflora* Roxb.
Sikakai	*Acacia concinna* (Willd.) DC.
Simuli	*Bombax ceiba* Linn.
Sirisa	*Albizia lebbeck* (L.) Benth.
Sisal fibre	*Agave sisalana* Perr. ex Engelm.
Sugandha/ Landa baguli	*Ocimum basilicum* Linn.
Sunari	*Cassia fistula* Linn.
Tala	*Borassus flabellifer* Linn.
Tentuli	*Tamarindus indica* L.
Versunga leaves	*Murraya koenigii* (L.) Spreng. Wall. ex Guill & Perr.

Note: N.A. = Not available; H = in Hindi

* The oil extracted from the seeds of the plants which are star (*) marked is also used for medicinal purposes.

Bibliography

Champion, H.G and Seth, S.K., 1968, *A revised survey of the forest types of India*, Delhi, Manager of Publications.

Cooperative Training College, Bhubaneswar, Orissa, 1989, *Background Papers and Recommendations of State Level Seminar on LAMPS* (Xerox copy).

Das N.K. and Routray Achyutananda, 1995, *The Orissa Forest Manual with Code, Orissa Law reviews*, High Court Square, Cuttack-753002, published by Smt. Susmita Rout, Printed at Mahavir Printers, Cuttack.

Fernandes, W. / Menon, G. / Viegas, Ph., 1988, *Forests, Environment and Tribal Economy. Deforestation, Impoverishment and Marginalisation in Orissa*, New Delhi, Indian Social Institute.

Government of India, Ministry of Home Affairs, 1979, *Marketing of Minor Forest Produce*, workshops held at the Administrative Staff College of India, Hyderabad 1977-79.

Government of India, Ministry of Home Affairs, New Delhi, 1980, *Report of the Working Group on Tribal Development during Sixth Plan*, 1980-85.Government of India Press, Faridabad 1981.

Government of India, Ministry of Home Affairs, New Delhi, 1984, *Report of the Working Group on Development of Scheduled Tribes during Seventh Five Year Plan*, 1985-90, Government of India Press, New Delhi, 1995.

Government of India, Ministry of Welfare, New Delhi, 1989, *Report of the Working Group on Development and Welfare of Scheduled Tribes during Eighth Five Year Plan*, 1990-95, Printed by Director, Institute for the Physically Handicapped, New Delhi, 1989.

Government of India, Ministry of Agriculture, 1995, National Oilseeds and Vegetable Oils Development Board, Background materials and proceedings of workshop on *Strategy for Development of Tree borne oilseeds and Niger in Tribal areas*, 20-30 August 1995, Bhubaneswar.

Government of Orissa, Forest Department, 1981, Chief Conservator of Forests, *Annual Administration Report of Forest Department*, Orissa for the year 1980-81.

Government of Orissa, Registrar, Cooperative Societies Orissa, Bhubaneswar, 1981, *Manual for Large Size Agricultural Multi Purpose Cooperative Society in Orissa (LAMPS)*.

Government of Orissa, Forest Department, Chief Conservator of Forests, 1983, *The statistical outline of Orissa Forests*.

Government of Orissa, Forest Department, Forest Utilisation Division Orissa, Cuttack, *Important Forest Species of Orissa yielding Non-Timber Products.*

Government of Orissa, Forest Department, Statistical branch, 1994, Office of the Principal Chief Conservator of Forests, Orissa, Bhubaneswar, *A Decade in Forestry in Orissa*, 1981-1990, Printed at Government of Orissa Press, Cuttack.

Government of Orissa, Harijan and Tribal Welfare Department, Bhubaneswar, 1990, *Tribes of Orissa*, Printed at Government Press, Cuttack.

Murdia, Ratna, 1990, Role of Forests in Tribal and Rural Development. Course material for the Training-cum-Workshop for *Senior Indian Forest Services Officers*, organised by Unit for Research and Consulting in Social Welfare Administration, Tata Institute of Social Science, Bombay and sponsored by Ministry Environment and Forests, Government of India (cyclostyle copy) 10-24 February 1990.

Nanda Amulya Ratna, Census of India, 1991 Series 1, India Final Population totals paper 1 of 1992, Vol. II, Census Commission, Ministry of Home Affairs, New Delhi.

National Bank for Agriculture and Rural Development, Bombay, 1990, *Study group on Lac — Report.*

National Bank for Agriculture and Rural Development, Regional Office, Bhubaneswar, 1995, *Development of Tribal Areas in Orissa — An Integrated Approach.*

National Bank for Agriculture and Rural Development, Bhubaneswar, 1995, *Economic Activities for Tribals in Mayurbhanj District, Orissa.*

National Bank for Agriculture and Rural Development, Bhubaneswar, 1995, *Horticulture and Tribal Development in Gajapati District, Orissa.*

Negi S.S., 1992, *Minor Forest Products*, Period Experts Book Agency, Vivek Vihar, New Delhi.

Orissa Forest Corporation Ltd., 1987, Souvenir, Silver Jubilee 1987, *National Seminar on Marketing of Rural Products.*

Orissa Rural Development and Marketing Society, 1995, Souvenir, *National Seminar on Marketing of Rural Products*, 27-28 October 1995, Bhubaneswar.

Padhy G.S., 1984, *Forest Resources of Orissa*, Published by Dr. S.S. Padhi, Printed at Udyog Printers, Unit-3, Bhubaneswar.

Patnaik N. (ed.), 1982, *Tribal Development Administration*, Organised by Tribal and Harijan Research-cum-Training Institute, Bhubaneswar (cyclostyle copy).

Roy Burman B.K., 1982, *Report on Committee on Forest and Tribals in India*, Ministry of Home Affairs, Tribal Development Division, Government of India.

Saxena, N.C., 1995, *Forests, People and Profit*, Dehra Dun, Natraj Publishers.

Thakur, D. / Thakur, D.N. (eds.), 1994, *Tribal settlement and minor forest produce.* (Tribal life in India, vol. 2), New Delhi, Deep & Deep Publications.

Visuals

1. Bamboo (kanta baunsa) *Bambusa arundinacea*

2. Scene of deforestation

3. Kapasia — *Kydia calycina*

4. Sidha — *Lagerstroemia parviflora*

5. Panas — *Artocarpus heterophyllus*

6. Sal — *Shorea robusta*

7. Barakoli — *Zizyphus mauritiana*

8. Sal — *Shorea robusta*

9. Dimiri — *Ficus glomerata*

10. Mahua — *Madhuca indica*

11. Patalagaruda — *Rauwolfia serpentina*

12. *Firmiana Colorata* — Bark is source of Fibre

13. *Ipomoea Sepiaria* — Climber for tying

14. *Solanum Sp.* — *Food plant*

15. *Colebrookia Oppositifolia* — used as avenue plantation a dominant shrubby element

16. *Vanda Tessellata* — An orchid plant

Index